AF574100

No. 2094
$15.95

BROWN'S SECOND ALCOHOL FUEL COOKBOOK

BY MICHAEL H. BROWN

TAB TAB BOOKS Inc.
BLUE RIDGE SUMMIT, PA. 17214

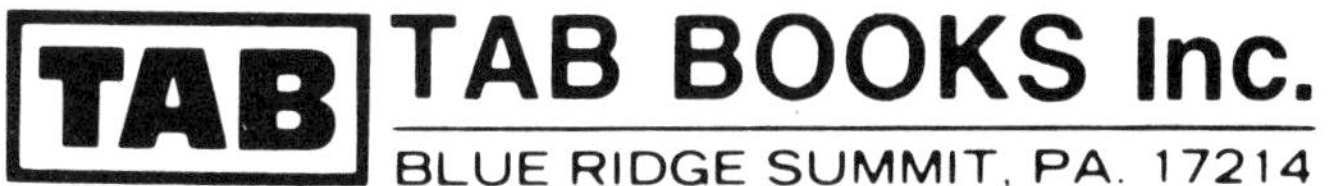

FIRST EDITION

FIRST PRINTING

Printed in the United States of America

Library of Congress Cataloging in Publication Data

Brown, Michael Halsey, 1942-
Brown's Second alcohol fuel cookbook.

Includes index.
1. Alcohol as fuel. I. Title.
TP358.B77 662'.669 AACR2
ISBN 0-8306-0048-5
ISBN 0-8306-2094-X (pbk.)

Cover courtesy of Industrial Safety Product News—an Ames Publication. Illustrated by Rodgers' Graphic.

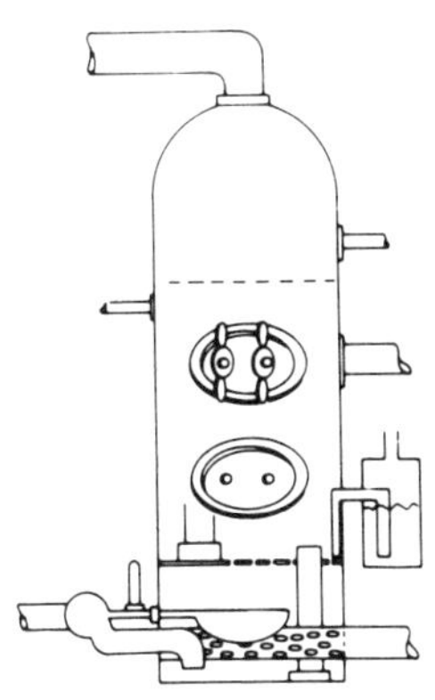

Contents

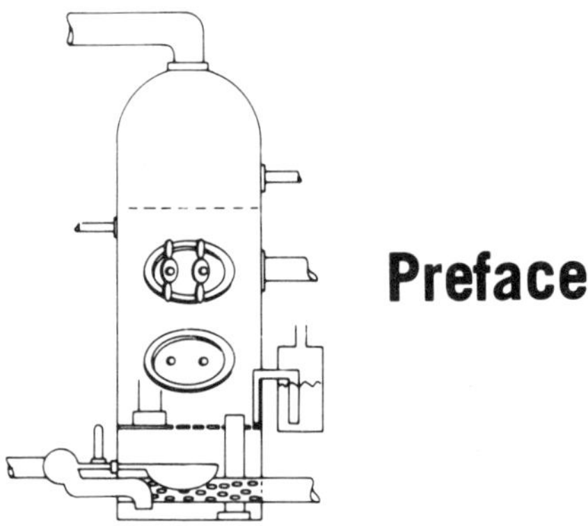

Preface

A lot of folks—especially farmers—have become extremely discouraged at the apparent lack of progress in developing motor fuel from farm crops and organic waste. As my grandmother used to remind me, if you want something done right—do it yourself.

This book will enable you to do it yourself. If the rest of the world wants to walk or drive around in cars the size of a postage stamp, that doesn't mean you have to join them.

I will teach you how to build a commercial-type still, make diesel fuel from soybeans, make motor fuel quality alcohol without using a still, and much more. I've done all the foregoing in the laboratory. It's easier than you think.

I will also teach you the proper ways to modify carburetors so that they will run alcohol and how to build dual-fuel units that—with the flip of a switch—will convert your engine fuel source from gasoline to alcohol while you are 55-ing it down the interstate. There are sections on drying the leftover mash, turning carbon dioxide into dry ice, and more.

And if you don't think we're making progress, read the testimony from the 95th Congress. I have included it as Appendix B.

Michael H. Brown

1

Building A Column

Everyone and his brother are apparently trying to get into the still-building business. On my last round of seminars through the Midwest, I saw some of the most amateurish junk imaginable being palmed off on the unsuspecting as "column stills." This included such things as pipe filled with metal scraps. In some cases, people who (apparently) know what they're doing won't let you have a set of plans until after you have purchased a complete distillery from them.

If you want to build a column still, maybe I can remove some of the mystery from it. If my approach seems a little over-simplified bear with me. If my directions seem so simple a child could understand them and you feel insulted, don't. I'm merely trying to protect myself from "free department" engineering requests over the telephone.

Start with the overall column. It is merely a long pipe (temporarily) sealed at both ends. See Fig. 1-1. There are more efficient ways to build a column than starting off with a pipe sealed at both ends, but I believe this is the simplest way to teach a concept.

THREE SECTIONS

Next divide this column into sections. The three main sections will be the *steam chamber,* the *stripper plate section*, and the *rectifying section*. It's put together as shown in Fig. 1-2.

In case you're wondering where the second column you see in all the ads went, that's the rectifying section. Most of the stuff you

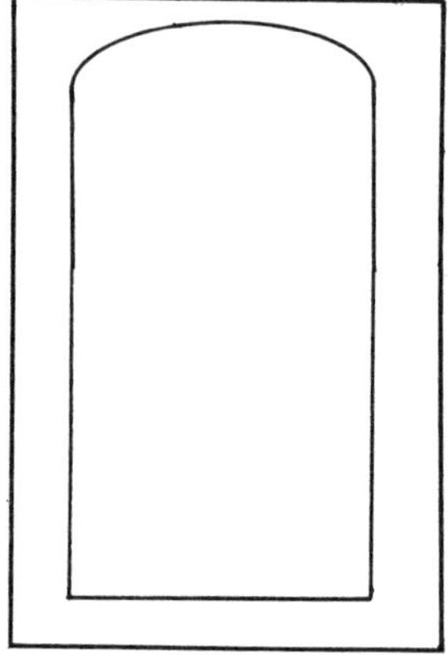

Fig. 1-1. The overall column.

see in the magazines consists of a steam chamber and stripper plate section in the first column with the rectifying section in the second column. In the distillery business, the second column is known as an *alcohol column* because the object is to get 190-proof alcohol out of it, 160-proof alcohol and below, technically, is whiskey. However, if I lay out a double column set-up here, it would be necessary to go into reflux lines and other subjects that might get confusing this early in the game.

With the column split up into the three main sections, add the pipes (or lines) going to and from the column. Ignore the reflux lines because that requires a separate explanation. Concentrate on the main lines only, the steam line to the steam chamber, the slurry line away from the steam chamber, the beer feed line to the

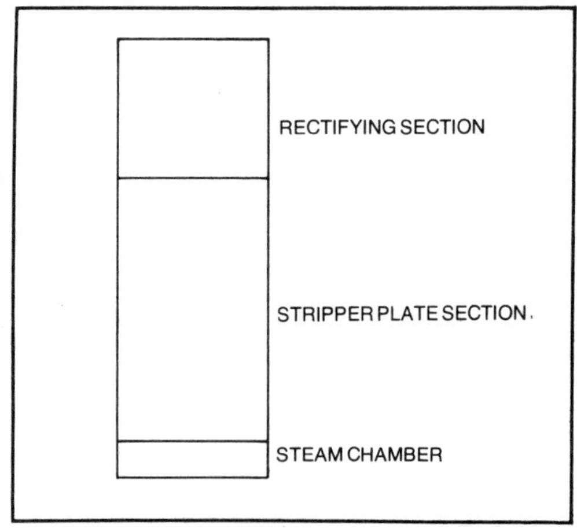

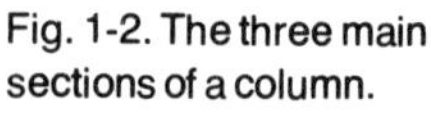

Fig. 1-2. The three main sections of a column.

stripper plate section, and the vapor line away from the rectifying section. Your lines will appear as shown in Fig. 1-3.

The way it works is that steam from a steam boiler is pumped into the steam chamber. Initially, the slurry line remains closed. You will see how and why in a bit. The superheated steam, 550 F at 1¾ pounds pressure, rises through the column and heats it up. When the column surpasses the boiling point of alcohol, normally determined by a thermometer plugged in halfway up the column, beer is pumped in from the mash tube and allowed to fall through the stripper plate section. Normally, you will have 16 to 23 stripper plates. See Figs. 1-4A, 1-4B, 1-4C and 1-4D.

BEER FEEDING

When the beer is fed onto the second stripper plate down (feed it onto the first plate and you stand a good chance of clogging your rectifying column), it consists of three materials: alcohol, water, and slurry. You want the water and slurry to go down into the steam chamber and the alcohol up into the rectifying section. Due to the laws of chemistry, you will be doing good if you get half alcohol and half water out of the top of the rectifying column. You get the proof

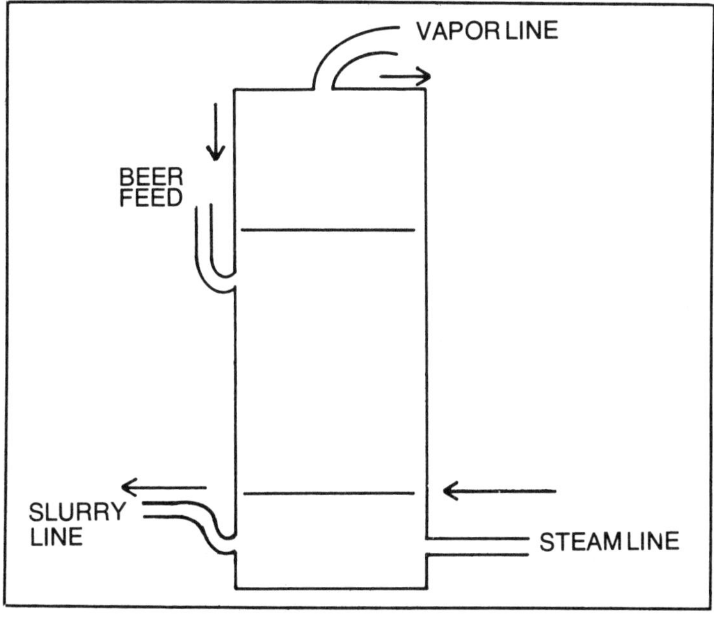

Fig. 1-3. The main lines.

Fig. 1-4A. Stripper plates. Holes should be 13/32″ in diameter and comprise 15 percent of the surface area of the plate.

Fig. 1-4B. Stripper column internals. A threaded rod allows the stripper plates to be moved up and down for adjustment. Nuts on each side of the plate hold them in place. Caps on the downpipes are for protection during shipping. The downpipe entrance must be lower than the liquid seal to make it work.

to 190 by refluxing. This is explained later. If your beer was 10 percent to start with don't get greedy.

Once the beer hits the second stripper plate, here's what happens. As the beer (or mash) drops down through the column (Fig. 1-5), steam from the steam chamber rises up through the column. The mash flows down through the *downcomer* (or downpipe), fills up the cup known as the liquid seal, and then flows onto the stripper plate. Mash travel is represented by the straight arrows in Fig. 1-6. Steam travels up through the holes on the stripper plate. At each stage, or plate, the steam heats the alcohol to its boiling point and carries it upward, or "strips" the alcohol from the plate. Hence, "stripper plate." Alcohol and steam travel is represented by the curved arrows in Fig. 1-6. This process is

Fig. 1-4C. A stripper column used for demonstrations.

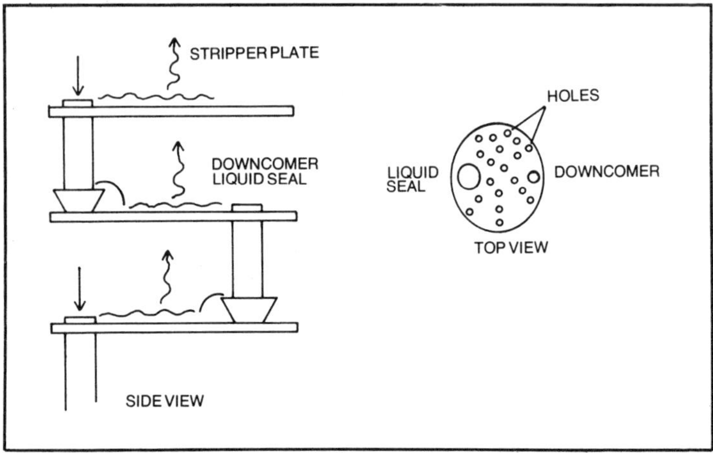

Fig. 1-4D. Stripper plates.

repeated until all the alcohol has been stripped from the mash and the slurry winds up in the steam chamber.

A word of caution to would-be buyers of distillery equipment. A lot of columns represented as "10 gal. hour" or "20 gal. hour" production rate are based on maximum feed rate. That means you will only strip out 1½ gallons of alcohol per bushel of corn instead of the 2½ gallons it should yield.

Pump the mash in too fast and almost half your alcohol gets pumped out with your slurry (or what you would *think* was spent mash). You can increase the number of gallons of 190-proof that your column puts out per hour this way, but it's extremely wasteful.

THE STEAM CHAMBER

Before going any further, I want to cover the steam chamber. There are three major parts in it; the *steam line,* the *float,* and the *slurry line*. Let's start with the steam line (Fig. 1-6). All there is to it is a pipe with oblong holes in it from the steam boiler to the far side of the steam chamber. The holes are merely to distribute the steam evenly through the falling liquid.

The slurry line is a must or you will have a column that simply fills up with water and distills nothing. See Fig. 1-7. On a full size commercial column, the top of the slurry line (T) and the steam line are usually parallel to each other. On small homemade jobs the setup is more like what is shown in Fig. 1-8. The elbow in the

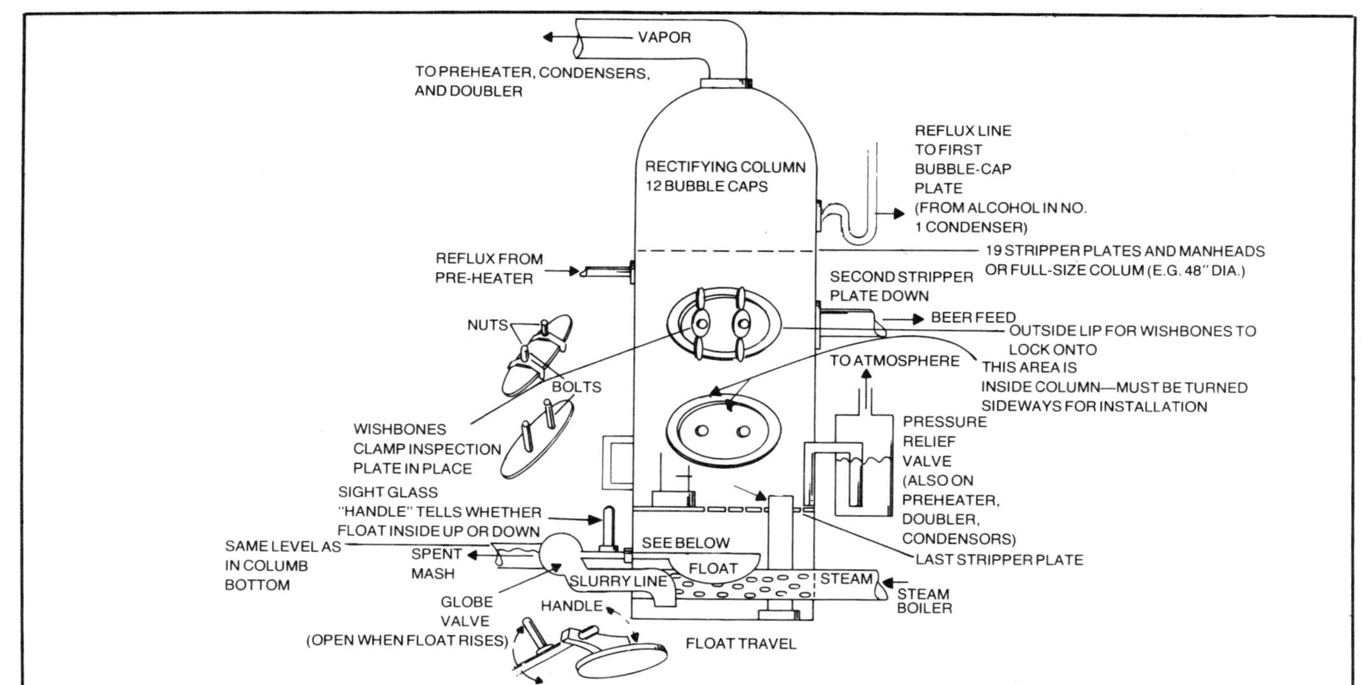

Fig. 1-5. The column.

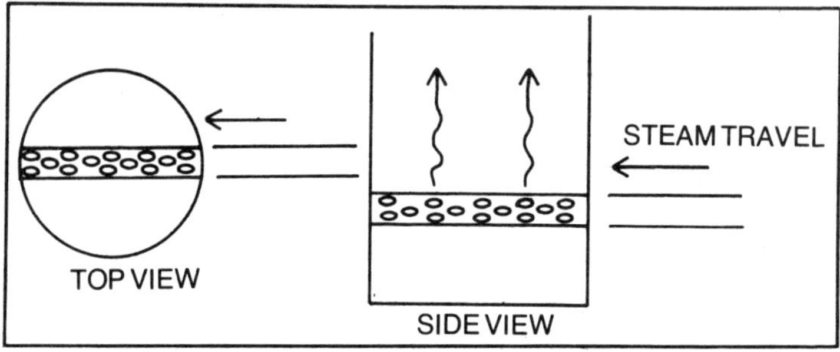

Fig. 1-6. Mash and steam travel.

slurry line acts as a liquid seal to keep excess heat from escaping from the steam chamber.

On a commercial or automated column, you will want to have a float (Fig. 1-9) and butterfly valve connected to your slurry line. They operate much the way the contraption in your toilet tank does to keep the water at a given level. In this case, when the slurry reaches a certain level, you want it to leave automatically via the slurry line.

The only function the float has is to rotate the bar it is attached to. The bar, in turn, is connected to a valve in the slurry line (usually a butterfly valve). When the bar is rotated clockwise (in this case anyway), the valve closes. When it is rotated counterclockwise, it opens. Water pressure pushes the float up to rotate the bar counterclockwise and when the water level drops the weight of the float rotates the bar clockwise. The whole steam chamber winds up looking like Fig. 1-5. In real life, most of this stuff goes in at an angle and it is not as sanitary looking.

The dotted line in Fig. 1-10 represents the bottom stripper plate. The last downpipe and liquid seal should be obvious. If you want to know what's going on inside your steam chamber at all

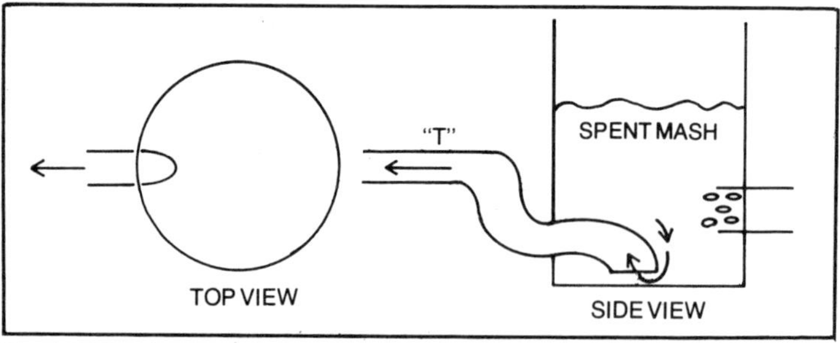

Fig. 1-7. The slurry line.

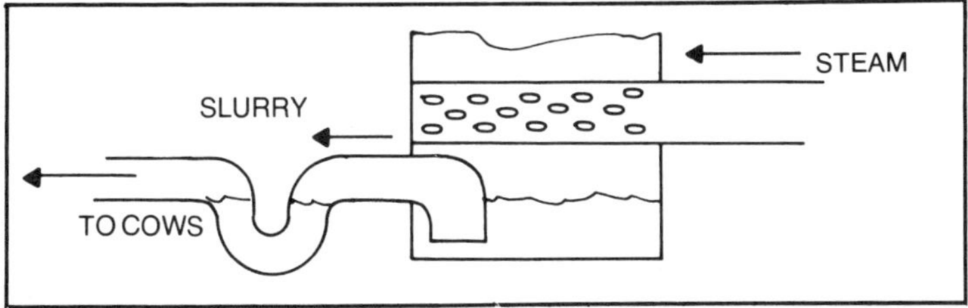

Fig. 1-8. A small column.

times, it is relatively easy to weld on another bar on the bar connecting the float and the slurry line valve on the outside of the column. When the float is as high as it will go and the valve is open, the bar should be sticking straight up. As the valve closes, the bar should "droop." See Fig. 1-11.

CONVENIENCE ITEMS

There are a couple of other convenience items that, while not absolutely necessary to the function of a column, might save you a lot of grief later on. One is a sight glass. It goes on the bottom stripper plate section on the outside of the column. See Figs. 1-12A and 1-12B.

If the water is at the levels indicated by the arrows inside the column, then the water in the sight glass will be at that same level. It is not uncommon for a column without a sight glass at that level to fill completely up with water before anybody figures out that something isn't kosher.

The other nice touch is a pressure relief valve. The idea here is that it is much more economical to blow excess pressure into the atmosphere than it is to send your column into orbit. A pressure relief valve will look something like the one in Fig. 1-13.

A pipe goes from about an inch above the bottom stripper plate into the pressure relief valve Fig. 1-14. A sight glass on the side of the valve indicates the water level in the valve (and valve only).

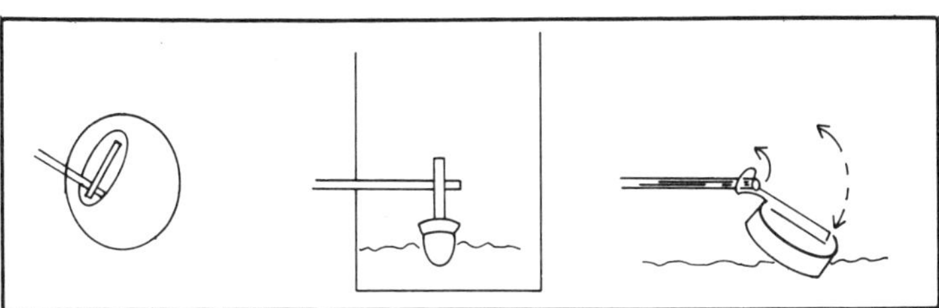
Fig. 1-9. A float.

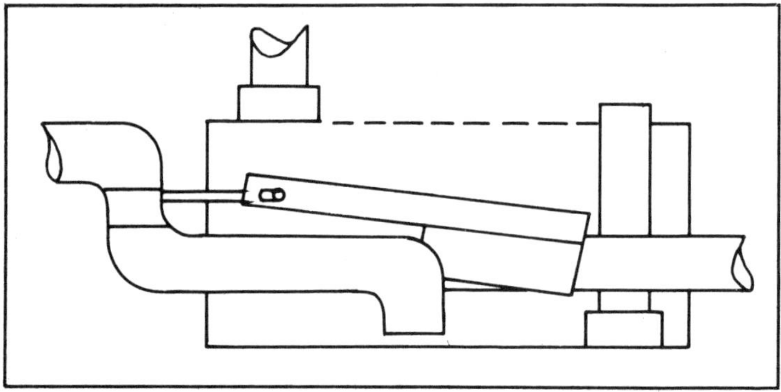

Fig. 1-10. The steam chamber.

The pressure of the water in the valve keeps normal steam pressure and heat from escaping. However, once the steam pressure exceeds a certain point the water pressure will no longer contain it and that steam then forces itself past the water in the pressure relief valve. Normally, the steam is vented outside of the distillery building well off the ground so that no one accidentally gets hit in the snout with it. Mounting on the column and venting arrangements are as shown in Fig. 1-15.

The last section of the column internals is the rectifying section. It consists of wine plates, bubble caps, and a stabilizing section. In the 19th century, this consisted of 12 sections with the vapor and reflux lines connected to a type of condenser called a "goose." Today, we use reflux pumps and an alcohol column

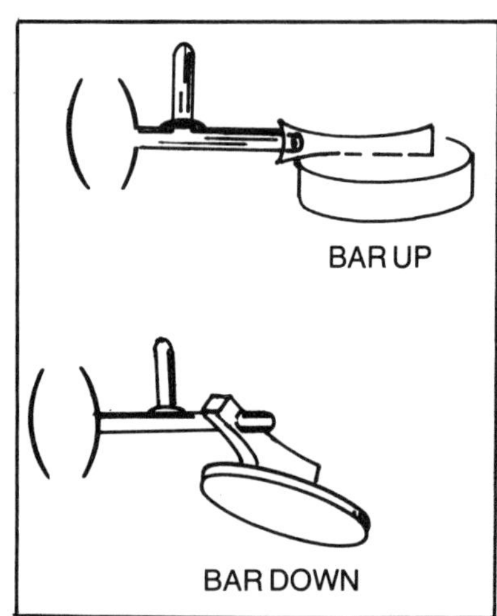

Fig. 1-11. The bar should "droop."

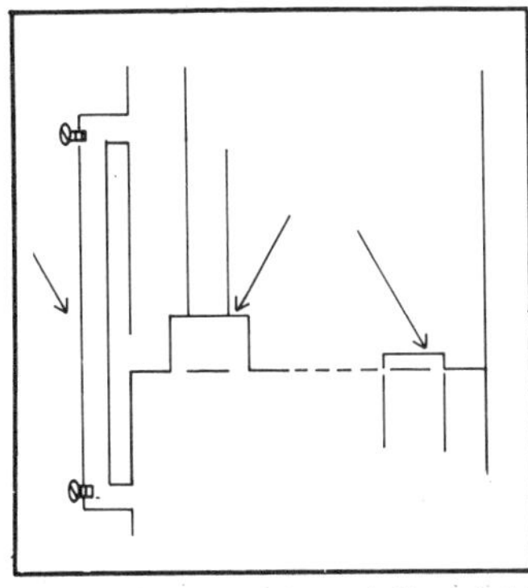

Fig. 1-12A. Location of the sight glass.

Fig. 1-12B. A sight glass at the base of a column.

designed for 190 proof that might have as many as 50 sections. The sections look like Fig. 1-16.

The way this section works is simply a refinement of the stripper plate section. Once the alcohol has been stripped from the mash the concentration of it must be increased.

The steam and alcohol vapors ascend through a series of pipes welded onto the wine plates (see A of Fig. 1-16). A slotted cap (called a bubble cap) fits over the pipe. Water level on each plate is

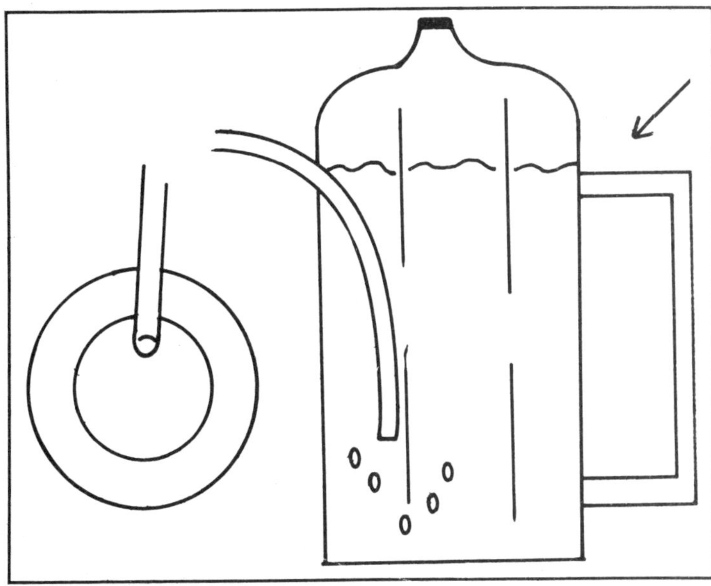

Fig. 1-13. A pressure relief valve.

Fig. 1-14. A blow-off valve for a stripper column. Note the sight glass on the right.

controlled by the pipe (B of Fig. 1-16) on the left side of the plate. When the water level reaches the uppermost opening of the pipe, it automatically flows down onto the next plate and eventually all the way down to the steam chamber. The filling or "charging" of the column is normally done through a connection on the top of the column. Draining is done either by a series of faucet-style valves or by a "weep hole" (C of Fig. 1-16). The weep hole is too small to

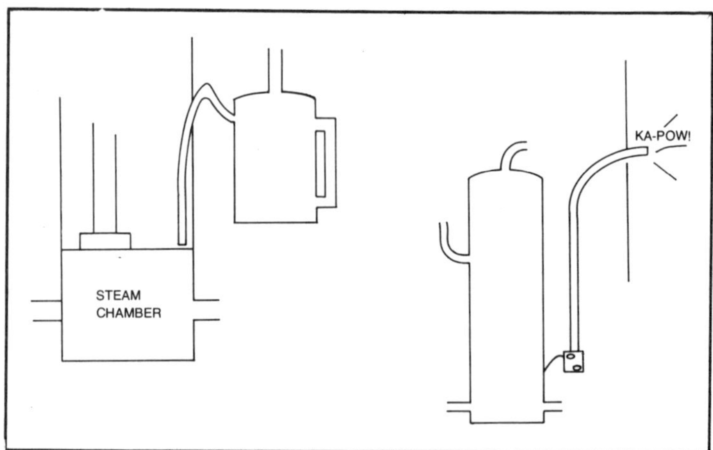

Fig. 1-15. Mounting and venting arrangements.

affect the operation of the column at half throttle, but it is large enough to let the column drain overnight.

As the steam and alcohol vapors flow through each stage (out through the slots in the bubble cap) more water and less alcohol is left behind at each stage. By stage 12, you should have about half alcohol and half water (100 proof).

Normally, the section second from the top will have a water level on the wine plate three to four times as deep as the other sections. This is known as a *stabilizing section*. It simply irons out fluctuations in beer feed rate, steam pressure and feed rate, and proof concentration.

No column ever works right the first time it is put together and operated. The plates are always the wrong distance apart. Figuring it out using chemistry and differential calculus is so complicated that even the chemical engineers go bananas with it. It's a lot easier to simply put a shaft down the middle of the column to hang the plate on and move them closer together and farther apart until you get it right. The farther apart your plates are (up to a point) the faster your mash feed can be, but the higher your column (or columns).

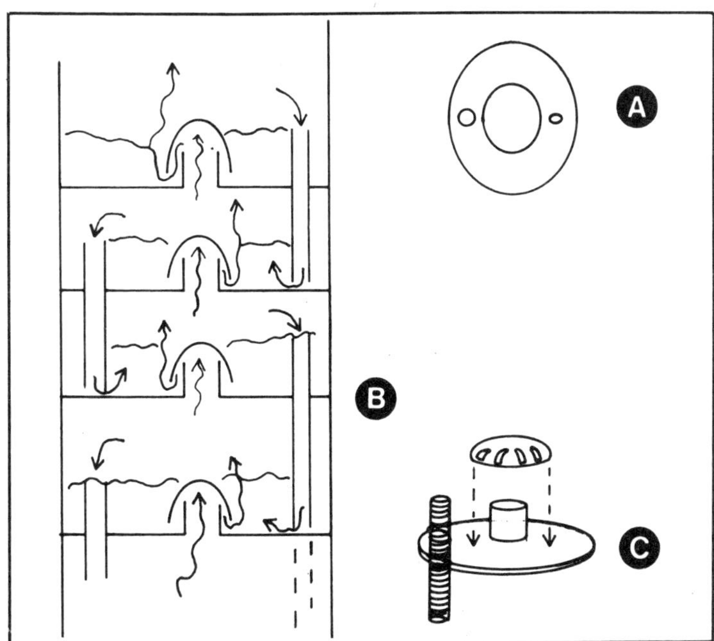

Fig. 1-16. The rectifying section.

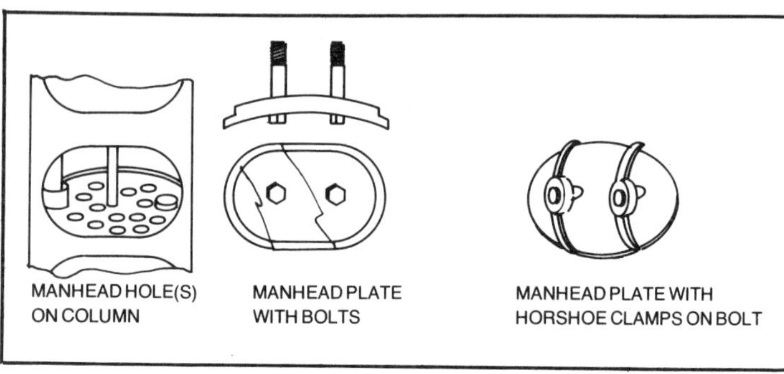

Fig. 1-17A. A manhead.

The wine plate is illustrated (Fig. 1-16) with one bubble cap for the sake of clarity. Commercial plates will have them sticking up all over the place to where they look like a series of mushroom farms.

A small column (e.g., a foot in diameter) is disassembled for cleaning. A large column usually has a *manhead* in front of each stripper plate to facilitate striking a hose in and washing it down. A rectifying section or column can go for months (or even years) without being cleaned.

A manhead has a lip that fits inside the column, two bolts through the manhead, and two horseshoe lockdown clamps (Fig. 1-17A and 1-17B). Notice the slight lip surrounding the stripper plate on the extreme left. This saves you from having to make a precision fit anywhere in this section.

Fig. 1-17B. A manhead on a stripper column. The box on the right goes to the instrument panel to record steam pressure.

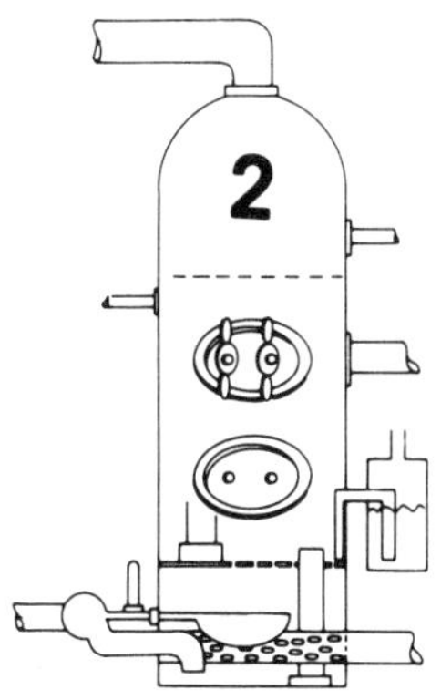

Stripper Columns For Sugar Substances

Stripper columns for substances such as molasses are considerably different from those that strip alcohol from starch. In a stripper column for corn or other starch, the downpipes and liquid seals are absolutely necessary pieces of equipment. On a molasses column they are not necessary.

The reason I use molasses as an example is that it is a fairly common substance used for the production of ethyl alcohol. You could just as easily go to a fruit cannery and ask for leftover fruit juice. The same rules would apply.

The construction of this column might seem a little strange to someone familiar with distillation equipment in this country. The pattern of construction I'm describing here is one that is commonly used in the Soviet Union.

Why the Soviet Union?

In this country, political and religious information—or misinformation, whatever your preferences—is cranked out in such volumes as to inundate even the unwilling. Getting technical information in the U.S.A. is harder than pulling teeth. Every company has closely-guarded trade secrets, labeled "proprietary information," that are kept from public scrutiny lest a computer latch onto something, use it to compete with the company that developed it, and cut into that company's profits.

In the Soviet Union there are no competing companies.

Consequently, there is no need to hide technical information. How-to-do-it books and articles on advanced chemical and other processes fall like rain over there. They're good folks to research

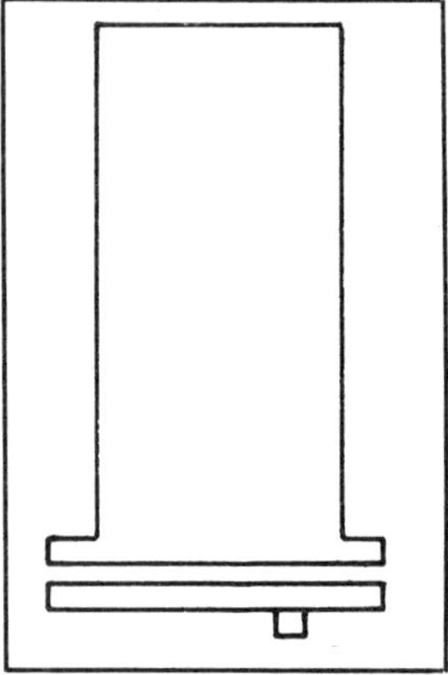

Fig. 2-1. A hollow cylinder.

from even though their form of government leaves a lot to be desired.

CONSTRUCTION

Construct this column the same way you did the last one: step by step. There is no need to dwell on the differences in the unit. They will become apparent soon enough.

First, make or use a long cylinder that is hollow at both ends (Fig. 2-1). The bottom of the cylinder is flanged so that you can attach a bottom cover with a drain valve. The drain valve is merely for washing the column down from inside during cleaning.

The top of the column is also flanged to facilitate the installation of a tapered top cover and hood. A steam pipe for

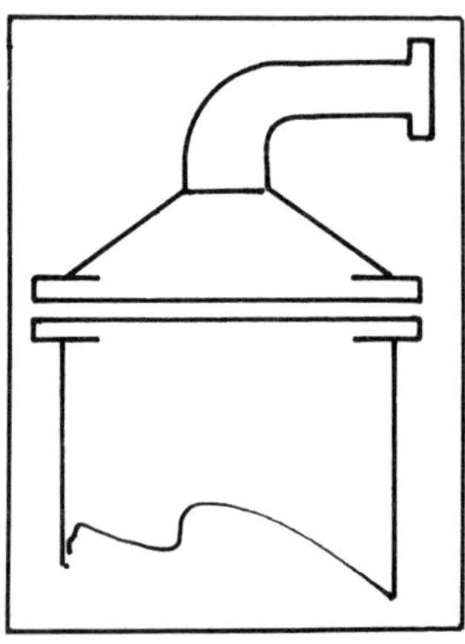

Fig. 2-2. The shaft is at the center of the column.

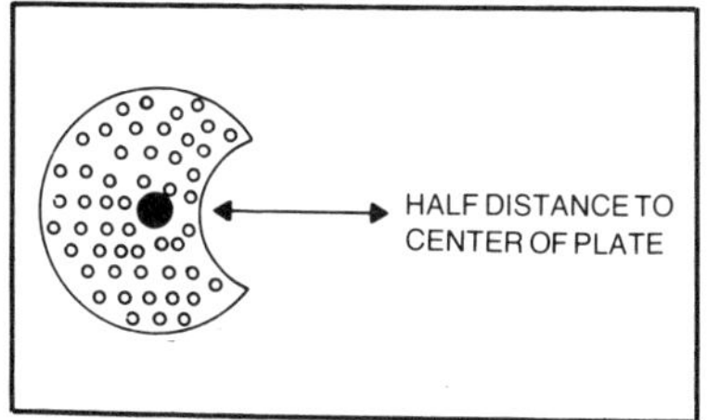

Fig. 2-3. Notches are cut in the stripper plates.

alcohol vapors joins the top cover/hood and goes to the rectifying column. That piece of equipment for increasing alcohol proof is identical in construction the world over.

A shaft runs down the center of the column (Fig. 2-2) to hang the stripper plates on. Because there are no downpipes for condensed water to return to the bottom of the column, crescent-shaped notches are cut in the stripper plates (Fig. 2-3). Saves a lot of plumbing, doesn't it?

The plates are staggered to keep a thin layer of water on the plates and to slow down the steam travel. The steam with alcohol vapor should stay inside the column no less than 15 to 20 seconds.

Obviously, the taller the column the more effective it will be in stripping alcohol and returning water. A table top model wouldn't work at all. The staggered plates from the side will look something like Fig. 2-4. Only eight plates are necessary in this type of column.

The steam chamber consists of three pipes that are mounted into the cylinder as shown in Fig. 2-5. The pipe on the left (Fig. 2-5) removes the molasses which has been watered down by steam. If they weren't watered down, they would gum up your column in short order. In a starch column, this would be equivalent to your slurry line.

The pipe on the top right (Fig. 2-5) is a steam line. Hot steam is introduced into the fermented molasses to break the alcohol and water molecules loose and send them up through the column. The

Fig. 2-4. Staggered plates.

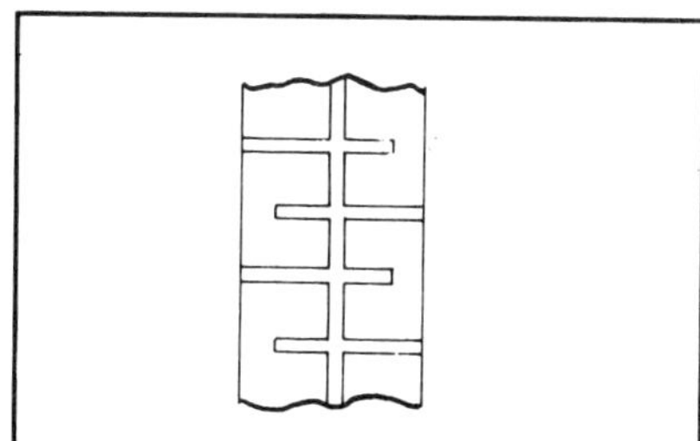

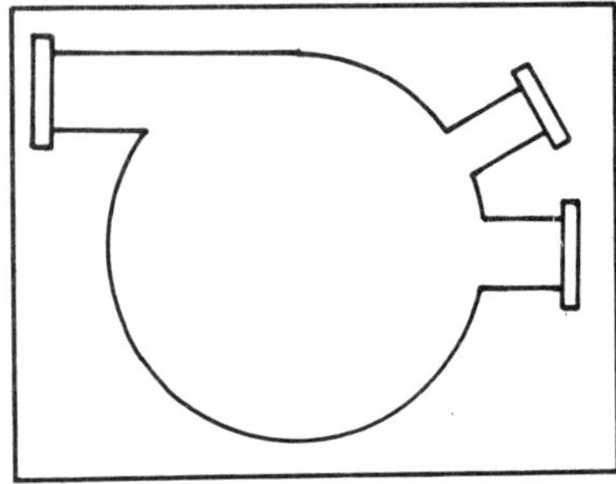

Fig. 2-5. The steam chamber.

pipe on the lower right (Fig. 2-5) introduces the molasses into the column. From a side view the pipes look like Fig. 2-6.

There is one more piece of equipment to add. Without it a tremendous amount of alcohol would simply escape. It attaches to the bottom plate and consists of a round flat iron plate with a hole in it, surrounded by what looks like a comic strip version of a crown. See Fig. 2-7.

This gizmo is removed by simply unbolting the bottom cover and yanking the whole mess out. The space between the bottom plate and the round flat iron plate surmounted by the crown is where the steam and fresh molasses are introduced—the vapors from which flow up through the hole.

Condensed water falls all around the iron ring on its flat surface courtesy of the notches cut in the stripper plates. When the water reaches the level of the bottom teeth of the crown, it begins to trickle back into the steam chamber, as opposed to having an overflow situation where the section overfills and a junior tidal wave takes place.

HOW IT WORKS

The entire operation occurs something like what is shown in Fig. 2-8. Grossly oversimplified, the alcohol and water vapor rises

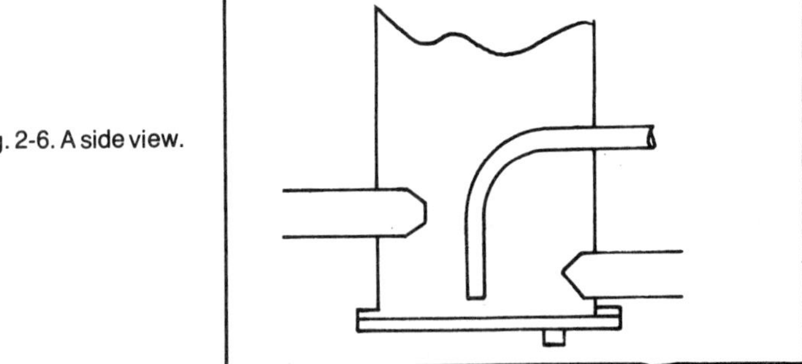

Fig. 2-6. A side view.

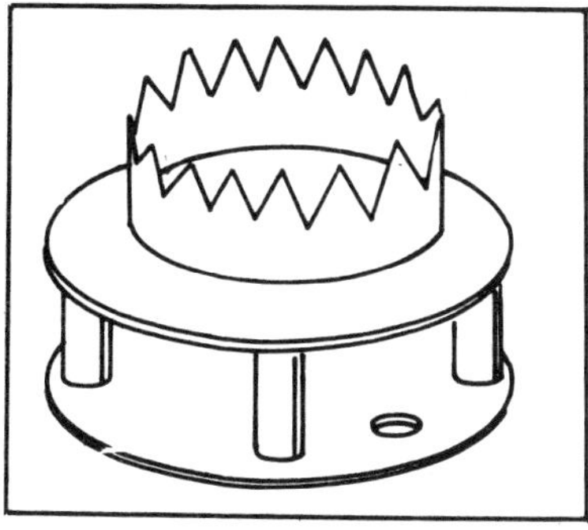

Fig. 2-7. The crown.

up through the hole and the water slides off the bottom stripper plate onto the flat iron ring and then out the overflow pipe on the left-hand side. Holes in the stripper plates can be smaller than 13/32 of an inch. No corn mash is going to get stuck in them.

Sizing the column, according to Soviet methods given in inches, is as follows. If the column is 15 feet tall it should be 3 feet in diameter, the molasses pipe should be 10 inches in diameter, the steam pipe 6 inches in diameter, and the overflow pipe 8 inches. A 10-foot column would be 2 feet in diameter with pipes scaled down accordingly.

In all honesty, I had to scale it down. Their idea of a small column is a 40-foot monstrosity.

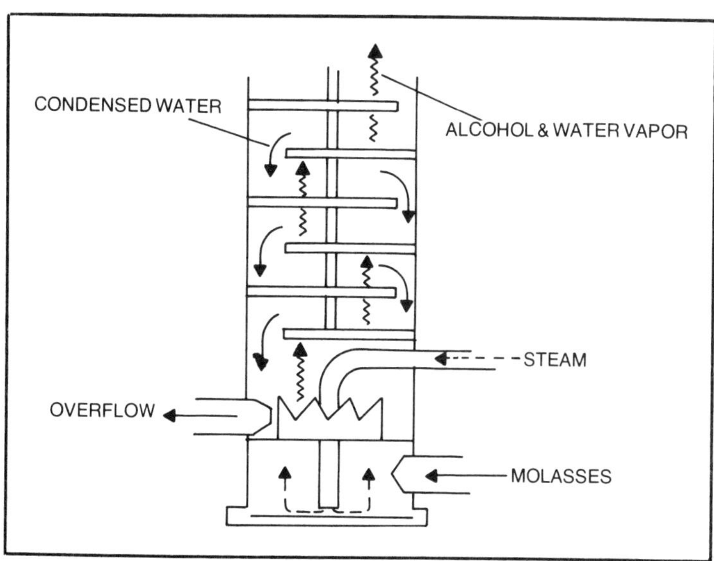

Fig. 2-8. The basic process.

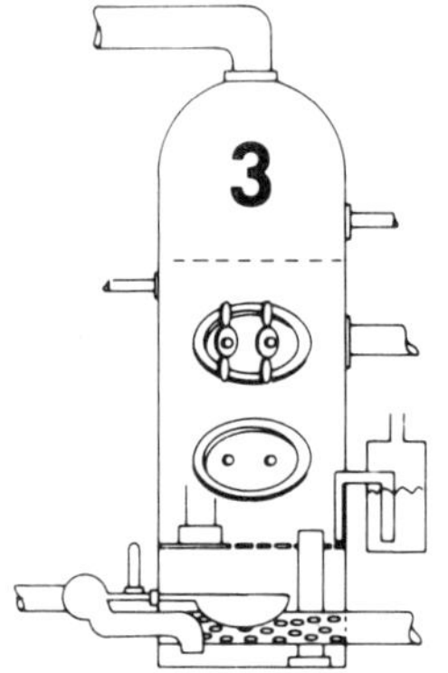

Reflux And Proof Concentration

The most difficult part of producing alcohol for motor fuel lies in raising the proof or concentration high enough to function satisfactorily in what we think of as a gasoline engine. A simple pot still and a doubler will produce 140-proof alcohol. While 140-proof alcohol will get you from point A to point B—in the summer—it won't idle well. And the energy absorbed by the water in the combustion chamber is tremendous. For really efficient year-round operation on today's low compression (for alcohol, anything under 12 to 1) engines, you need 190-proof alcohol. Or close to it.

HIGH-PROOF ALCOHOL

There are two ways to get high-proof alcohol. You have to combine them in real life to get the job done, but on paper these methods can be separated to teach the concepts. One concept is the principle of *infinite stages*. If you have wine plates with bubble caps and downspouts stretching from here to eternity, you should be able to get 190-proof alcohol out of the top of it. For obvious reasons, this is impractical. It would be a sort of Tower of Babel project.

The other concept is the principle of *total reflux*. That is, if you take 100-proof alcohol off the top of your column and run every drop of it back into the column after it has condensed to a liquid, then you will eventually achieve 190-proof alcohol. The problem here is that the 190-proof alcohol will never leave the column.

Both plates (stages) and the feed lines (reflux) constitute the dictionary definition of reflux: to flow back or return. The

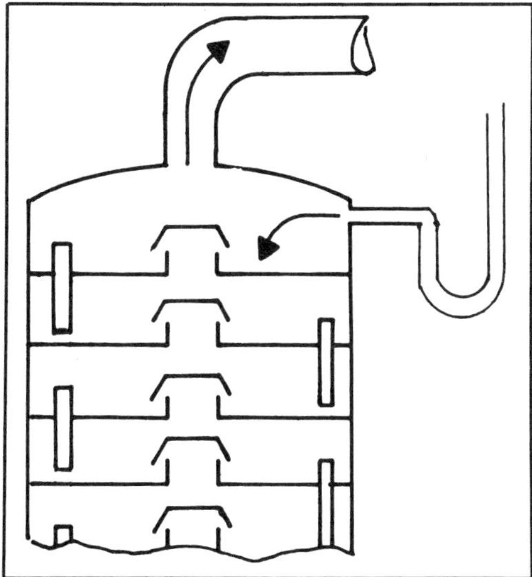

Fig. 3-1. Reflux lines.

bubblecaps and downpipes simply assist in vapor separation; the alcohol vapors rise and the water vapors rise, condense, and return to the bottom of the column. The reflux lines dump liquid enriched alcohol and water into a wine plate, a higher percentage of alcohol rises, the water rushes down the downpipe, and the cycle repeats itself. See Fig. 3-1.

The curved loop at the right (Fig. 3-1) represents a reflux line spilling liquid alcohol and water onto the first wine plate. The loop on the line furnishes a liquid seal from the condenser to the column. Otherwise, both arrows in the drawing would be going in the same direction. A schematic for a reflux system looks like the one in Fig. 3-2.

The arrow pointing to the left (Fig. 3-2) represents the reflux being returned to the top of the column. The arrow pointing to the right represents alcohol going to a storage tank. There is no condenser shown in this drawing.

A total reflux system would look like the one shown in Fig. 3-3. The mathematical formula for a reflux system is

$$R = \frac{L}{D}$$

Fig. 3-2. A schematic for a reflux system.

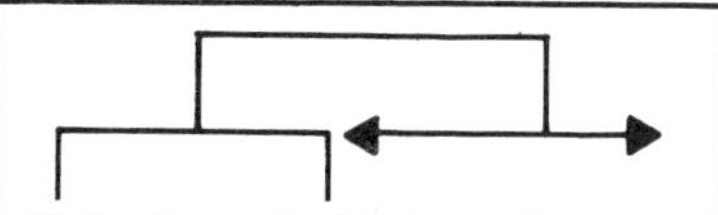

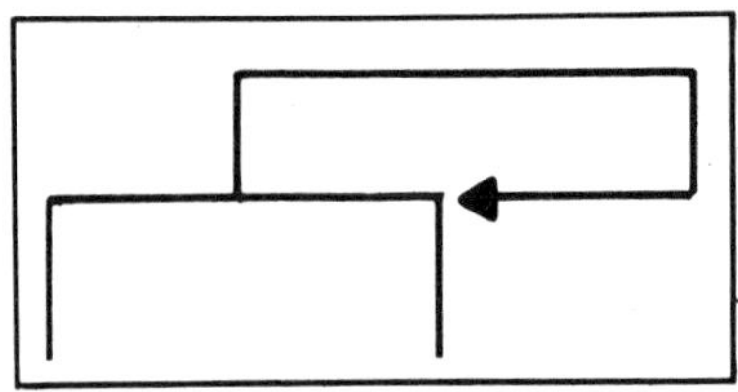

Fig. 3-3. The total reflux system outline.

where R is the reflux ratio and L is the flow rate of an overhead product.

In chemical engineering, this equation is in moles per hour. A mole is $6.02 + 10^{23}$ molecules. Let's ignore all this and do it by seat-of-the-pants engineering instead. More elbow grease is involved, but that's alright; half the time the engineers don't get it right either.

So how do we get it right?

Start with the stripper and rectifying columns in the previous chapter. Separate them this time into a stripper column and an alcohol column. In Fig. 3-4, the stripper column is on the left. Assuming a minimum of 16 stripper plates and a beer feed onto the second stripper plate from the top, you can safely assume that you will strip all the alcohol from the mash if the beer is not pumped in too quickly.

Where you will encounter the need for the first reflux line is in the alcohol column (the one on the right in Fig. 3-4). The excess water in the bottom of the alcohol column has to be put somewhere. Besides, some of the stuff might have alcohol in it.

The only thing to do is dump it back on the top plate of the stripper column. The vapor line from the top of the stripper column is entered onto the second wine plate from the bottom in the alcohol column. The bottom wine plate or chamber contains a reflux line leading to a pump. The pump then moves the reflux back into the stripper column. Normally this is to the third stripper plate from the top.

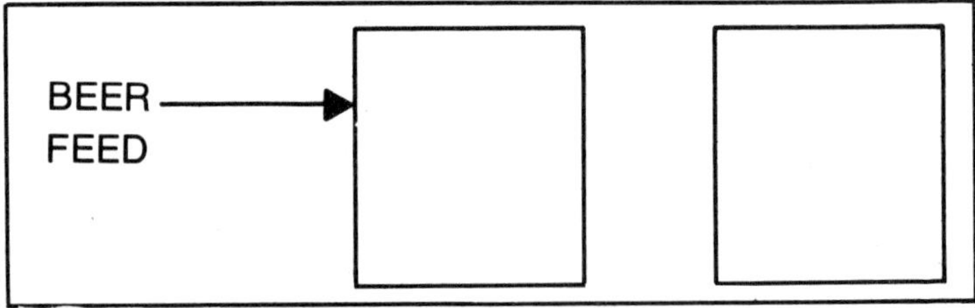

Fig. 3-4. A stripper column and an alcohol column.

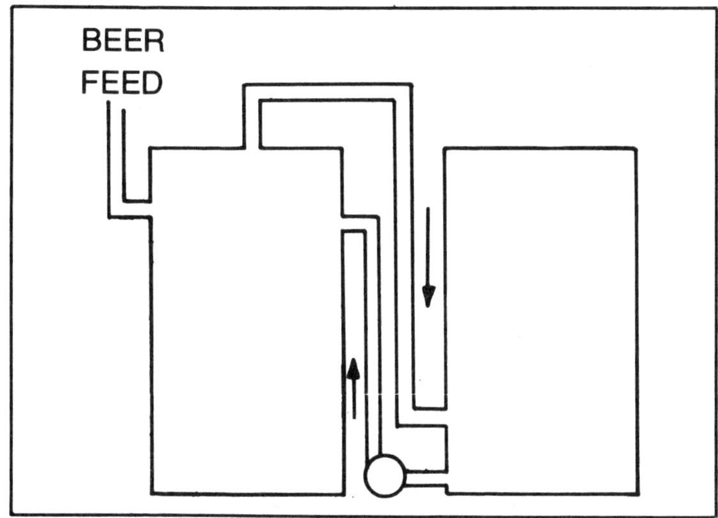

Fig. 3-5. A stripper column and an alcohol column.

In Fig. 3-5, the arrows going down represent vapor going to the second wine plate from the bottom. The arrow (Fig. 3-5) going up represents liquid reflux being pumped back into the stripper column. The best way to do this is to install an electric switch on the pump that will activate the pump every time a certain liquid level is reached at the bottom of the alcohol column.

Which brings us to the $64,000 question; how many plates do we need in the alcohol column?

As many as it takes.

This isn't as dumb as it sounds if you allow for the upper *pinch zone* or zone of constant composition. Let's say you start with a dozen wine plates. Let's say you're getting 90-proof alcohol out the top of your column. You add two plates. Now you get 95 proof. You add two more plates. Now you get 100 proof. You get up to 23 wine plates in your alcohol column and get 130 proof. But then you add seven more plates and still get 130 proof. Those last seven plates make up a "pinch zone" and are not required for anything except stabilizing your vapor flow. All seven (or however many you added)

Fig. 3-6. A reflux schematic.

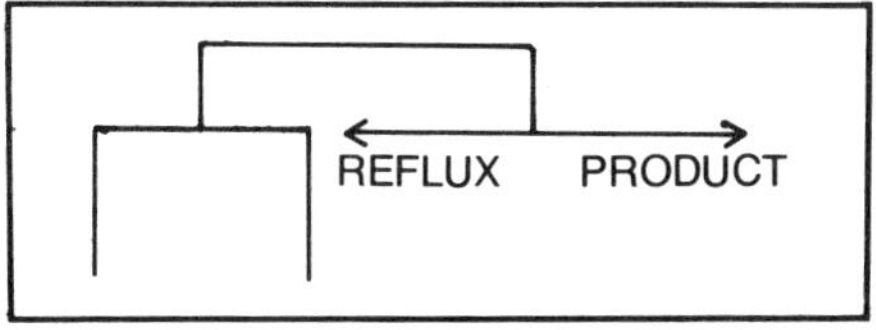

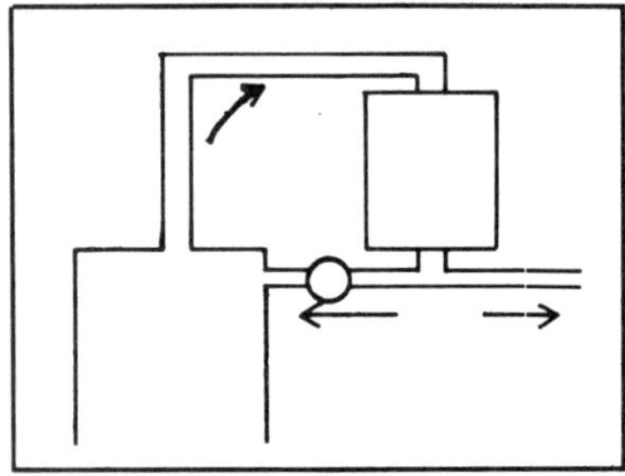

Fig. 3-7. An alcohol column with a reflux system.

can be replaced by one deep water section, about 9 inches of water on the plate, called a *stabilizer section*.

Most wine plates are within 6 inches to 9 inches of each other. This is true even on 5-foot wide commercial columns. The maximum number of wine plates that I have ever heard of is 50.

Once you have determined your zone constant composition, it is time to install your reflux systems. The reason you spend so much time on determining the maximum number of plates needed is that more plates (up to a point) equals less reflux. Less reflux equals less heat. Less heat equals less cost.

Back to our reflux schematic (Fig. 3-6). In real life, your alcohol column with a reflux system would probably look more like the one shown in Fig. 3-7. In the schematic the condenser and the pump are ignored. In this case, the speed of the pump determines the percentage of liquid reflux that is pumped back into the column.

There are several ways to handle reflux. At least one of the concepts goes all the way back to the early 19th century. The ideas presented here are by no means all inclusive. Feel free to use your own imagination and come up with improvements.

The first concept I will cover was once known as the "goose." Its technical name would label it a *dephlegmator*; a partial

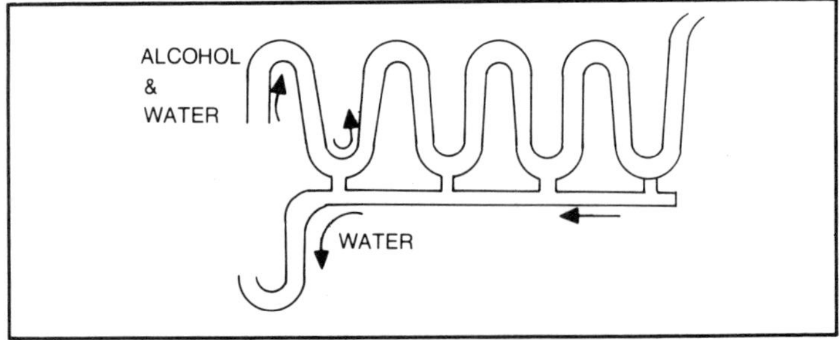

Fig. 3-8. The "goose."

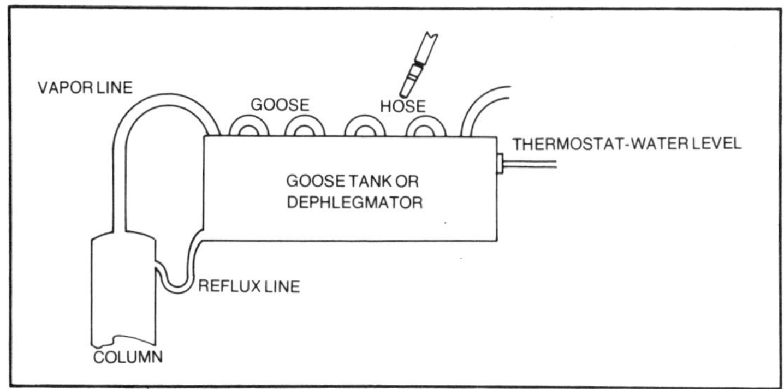

Fig. 3-9. A system for providing replacement water.

condenser to cool mixed vapor and, thus, condense the higher boiling portions.

What the "goose" resembles is a steam radiator with holes drilled at each bend in the bottom, a pipe connected to each bend, and a master pipe connected to all pipes at each bend serving as a drain and reflux lines (Fig. 3-8).

TEMPERATURE

The alcohol and water vapor leaves the top of the column and travels through the goose. The temperature of the goose is kept between 175° F. and 180° F. Because alcohol vaporizes at 173° F, it simply travels up and down the curved passages until it exits the device (right side of Fig. 3-8). However, most of the water cools below its boiling point of 212° F and slides down the loops and into the master pipe/reflux line.

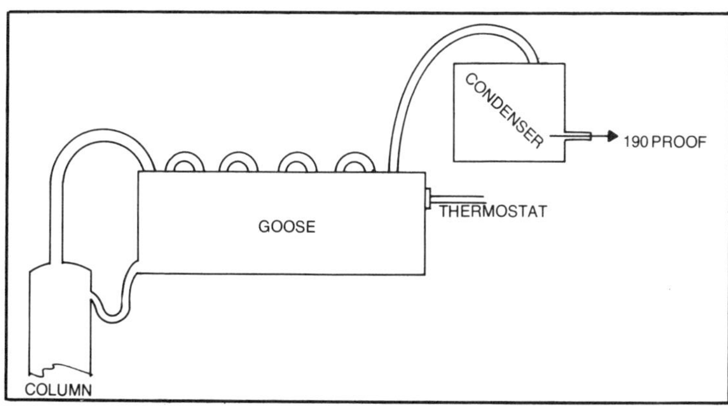

Fig. 3-10. A one-column system.

Fig. 3-11. A dual-pump system for a column.

The temperature of the goose is controlled by merely allowing the heat from the column to permeate the water in a tank holding the goose. An automobile thermostat at the water level of 175° F to 180° F rating will allow water that is too hot to flow out. A garden hose stuck in the goose tank can provide replacement water (Fig. 3-9).

The garden hose should be stuck all the way to the bottom of the tank because the hot water will rise to the top. If you are running a high volumn of alcohol, you will need more bends on the goose. You might want to pipe your overflow valve line to the water you are going to use for your steam boiler because the water is already warm and requires less energy to turn to steam.

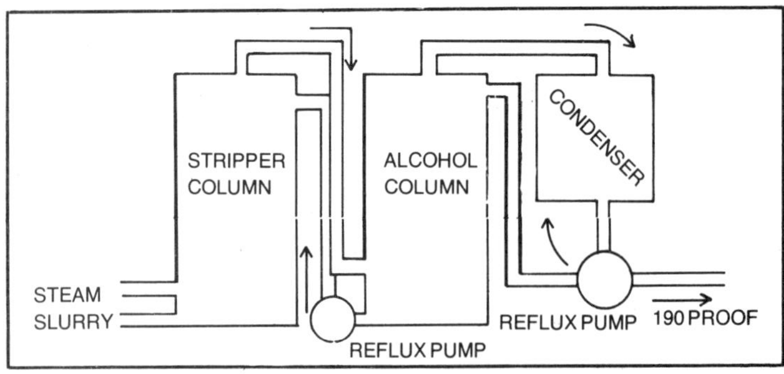

Fig. 3-12. A standard refluxing motor, pump and lines.

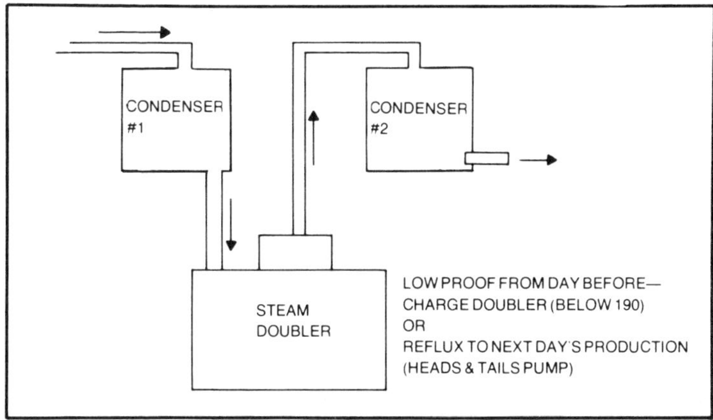

Fig. 3-13. A diagram of the reflux system.

MAXIMUM EFFICIENCY

For maximum efficiency with this device, the mixture coming off the top of your column should be slightly less than 100 proof. It will be 190 proof by the time you get it all the way through a properly constructed goose.

If the column is set up on the first floor of a building, the goose placed on the second floor, and the condenser on the third, reflux pumps can be dispensed with entirely. This is extremely handy if you are in Western Samoa, Darkest Arkansas or some other uncivilized places. Only a one-column system will work like this (without reflux pumps) unless you want to apply heat under both columns. See Fig. 3-10.

The most common procedure is to simply mount a centrifugal pump on the ground next to a column and simply pump a percentage

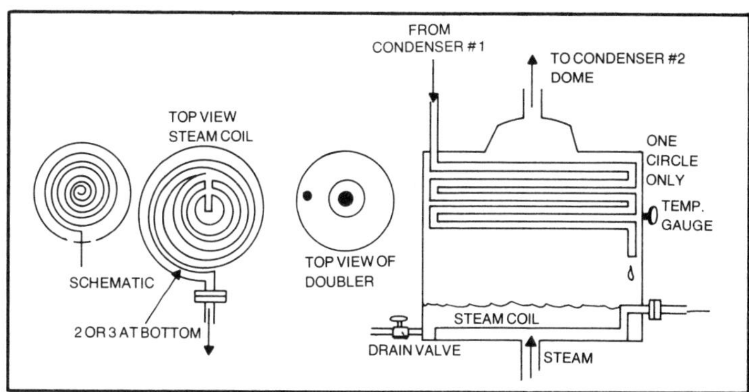

Fig. 3-14. The construction of the steam doubler.

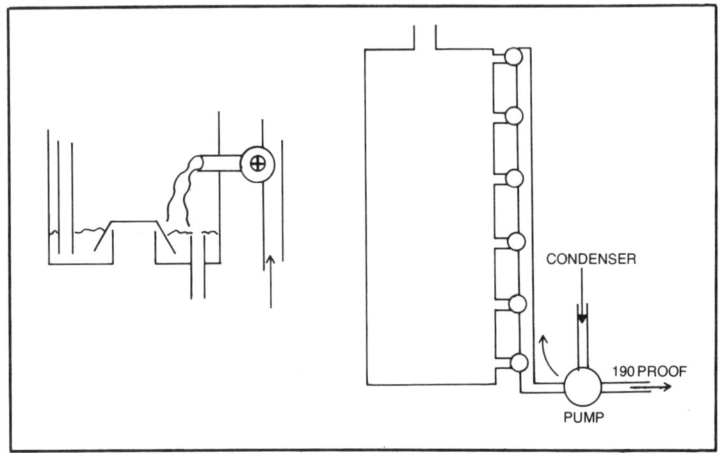

Fig. 3-15. A rectifying column with reflux valves.

of the condensed alcohol back into the top of the rectifying or alcohol column (Fig. 3-11). Easier said than done. If you want to reflux two-thirds of your product, that means that you have to pump 2 gallons back for every 3 gallons that comes off your condenser. Try it sometime. Without some sort of automatic controls, you immediately have large headaches. Some of these problems can be overcome with practice. A standard refluxing motor, pump, and lines will look something like Fig. 3-12.

There is normally another reflux line used if you have a preheater, but I will leave that for the chapter on preheaters and heat exchangers.

Fig. 3-16. Rectifying column with reflux valves.

STEAM-CONTROLLED DOUBLER

Another method of raising proof consists of using a steam-controlled doubler. This type of doubler eliminates much of the refluxing normally necessary and in turn saves energy and cuts cost.

Reflux from the condenser to the column is still necessary, but in much smaller amounts. What happens here is that you use two condensers (the first one is not a dephlegmator) with the steam doubler in between them. The outside of the setup is as shown in Fig. 3-13.

For the sake of simplicity reflux from condenser #1, drain valves, overflow valves, and so on are not shown in Fig. 3-13. The arrows indicate the flow of alcohol and water.

Condenser #1 turns the vapor to a liquid. It then enters the steam doubler, falling to the bottom. The temperature in the steam doubler is kept as near as possible to the boiling point of alcohol. At the beginning of a day's run, there might be only 6 inches of liquid in the bottom. The alcohol vaporizes and rises into condenser #2. Most of the water, the heaver boiling fraction, remains in the doubler. At night, the water level might be as high as 6 feet. When the operation is shut down the doubler is then drained.

The inside of the steam doubler is constructed as shown in Fig. 3-14. The line from condenser #1 is wrapped around the inside doubler wall to pre-heat the mixture before it is dumped into the bottom of the doubler. The steam is piped in through the bottom, circulated on the floor of the doubler to warm the chamber evenly, and evacuated out the side.

A temperature gauge is installed on the outside for internal reading of heat in the doubler. Keep the temperature of this thing

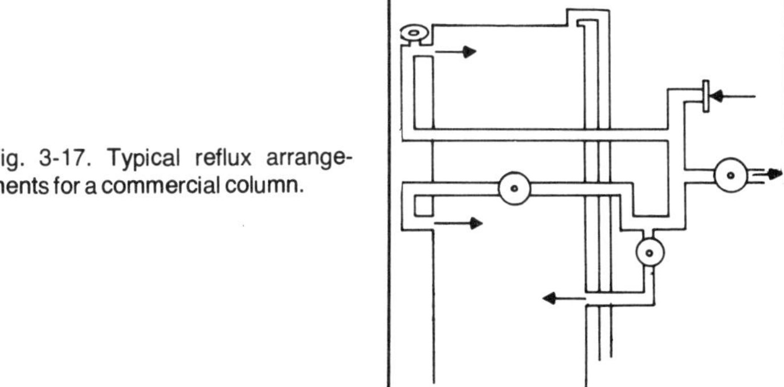

Fig. 3-17. Typical reflux arrangements for a commercial column.

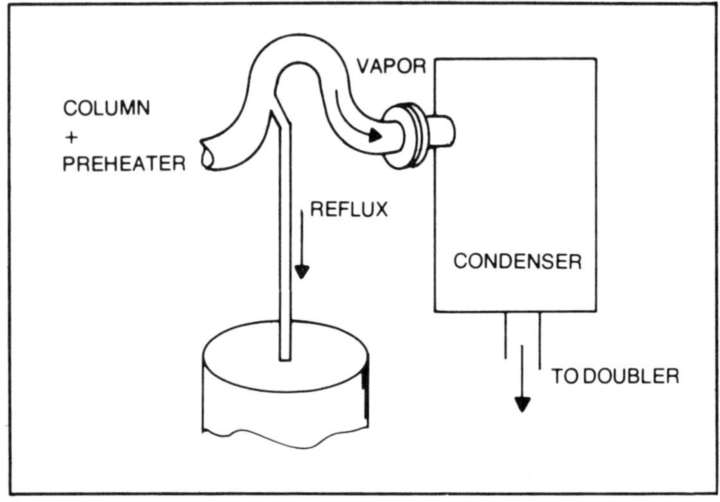

Fig. 3-18. A typical system with the control pot at the center.

over 200° F and you won't get over 160-proof alcohol no matter what you do. A drain valve is installed near the bottom and a sight glass is installed on the side.

Before I describe automatic reflux controls, there is one other item that bears mentioning. Running all your reflux to the top wine plate doesn't always get the job done.

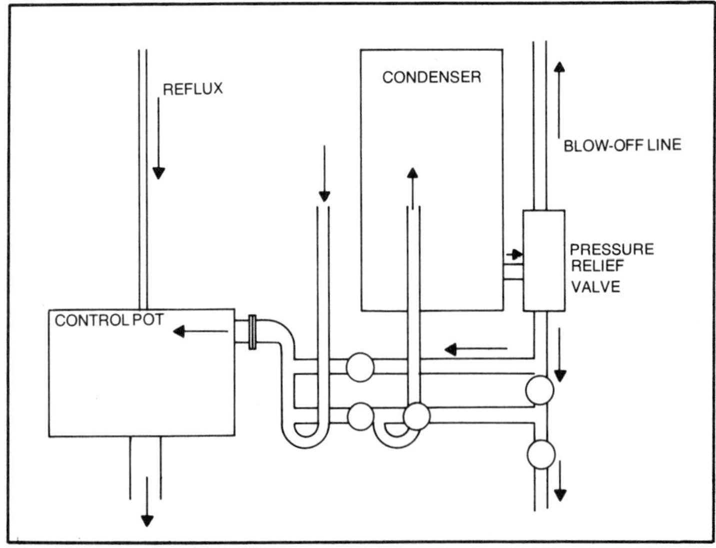

Fig. 3-19. Slightly more complicated reflux lines.

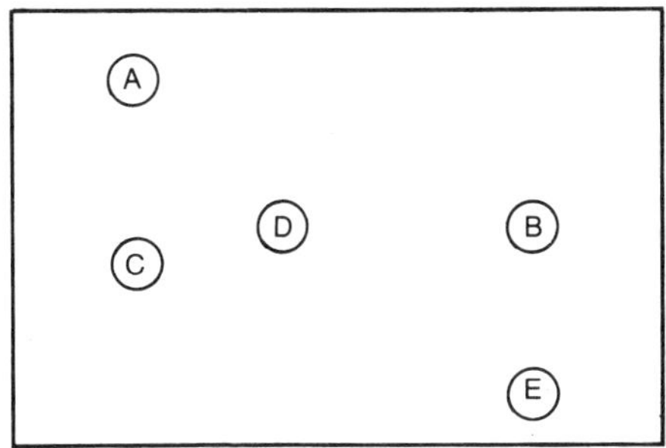

Fig. 3-20. A valve schematic.

You might have to run additional reflux to plates lower down to get the job done. The simplest (but not necessarily the cheapest) way to do it is to simply install a line of pipe fittings, valves, and master pipe alongside the column. Be sure the pipe nipples or fittings empty into the top of each wine plate chamber. See Figs. 3-15 and 3-16.

The inexpensive valves are normally the type you use to turn your garden hose off and on. Globe valves are less likely to leak and cause installation problems. Whatever wrinkles your prune.

This system can also be used as a clean-out line if another valve is installed and a hose run to it. This might create a problem in that when you fill the column up with water, the coarse particles left on the stripper plates float up into the bubble caps and clog them. When that happens, the whole column has to be taken apart.

On a huge commercial column, you might see all sorts of odd reflux arrangement. Figure 3-17 is typical. In this particular system, the top left reflux line empties onto the top wine plate, the bottom left line onto the bottom wine plate, and the bottom right line onto the third stripper plate from the top. The top right pipe

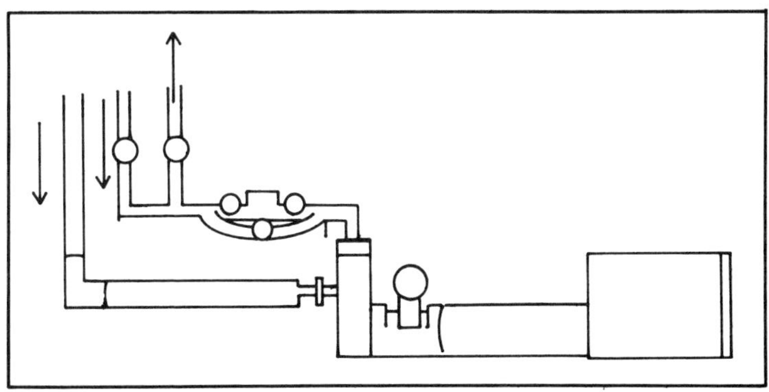

Fig. 3-21. The pump and lines.

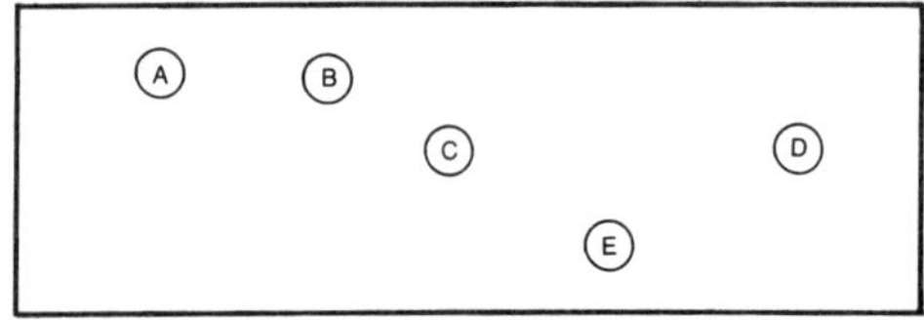

Fig. 3-22. Round valves.

with the flange is routed through the preheater and the bottom right open pipe—with the valve—if the valve is open, returns reflux to the control pot.

The pipe behind the reflux lines is simply, a clean-out line. In this case, it is a consistent clogger (with mash particles) of wine

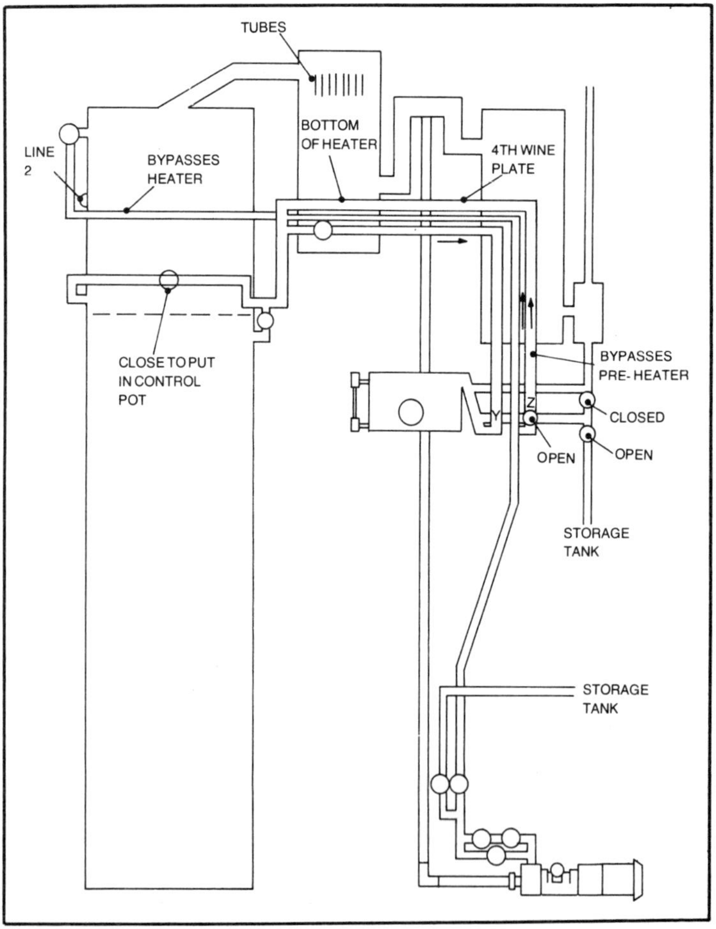

Fig. 3-23. The entire reflux system.

plates and bubble caps. Where the reflux is routed and in what amounts is simply a question of which valves were opened and how much.

In a whiskey distillery, you would also have a "heads and tails" pipe, storage tank, pump, and control box to aid in manipulating the flavor of the whiskey. That need not concern us.

AUTOMATIC CONTROL OF REFLUX

Automatic control of reflux is done with a *control pot* and it is usually part of a package installed in a large distillery. This is known as *full instrumentation*. Operations are done on the assumption that it is easier to install electronic and air compressor equipment to monitor distillery goings-on than it is to hire a bunch of potential alcoholics to watch the dials and pull the levers.

A typical system with the control pot as the center of attention is put together as shown in Fig. 3-18. What happens is that the hot

Fig. 3-24. The cylinder on the left is a preheater. The larger cylinder on the right is a rectifying column.

vapor rises off the column, loses some of its heat to the preheater (or beer heater). Some of the heat condenses in the inverted U-shaped pipe and runs down into the control pot. This is not the only reflux line going to the control pot; it is simply a handy place to put one.

The reflux lines on the other side of the condenser going to the control pot are a trifle more complicated (Fig. 3-19). The round objects on the pipes are valves. For this particular situation, make a valve schematic (Fig. 3-20) and label the valves A through E. Close off valves A and B and nothing will come out of the condenser. Open valve A and close the rest of them and everything goes in the pot.

Don't worry about where everything leads to at the moment. It will all match up later. Just get an idea of where everything goes and we'll put the entire package together at the end of the chapter.

The third part of the system is the reflux pump and controls—a sensor and an air regulator valve—on the control pot. In a full-size distillery, the reflux pump will be on the first floor, the control pot on the second, and the upper reflux lines on the third. The column, from top to bottom, fills all three floors.

The pump and the lines to it looks something like those shown in Fig. 3-21. the round valves on this diagram are as shown in Fig. 3-22. Valve A is on the same line as valve E (in Fig. 3-21). Valve B goes to the top wine plate on the column. E is a bypass valve that should be opened when either C or D malfunction. The square gizmo between C and D is an automatic air-operated valve. The parts of the reflux pump are, from left to right (Fig. 3-20), a centrifugal pump, bearing with chain oiler, shaft, and electric motor.

When the liquid falls in the control pot, a sensor triggers the electric motor on the reflux pump. An air regulator valve on the control pot prevents the liquid from rising over a certain level. A sight glass on the side of the control pot tells the operator the liquid level.

The entire reflux system I have just described will look something like Figs. 3-23 and 3-24. In Fig. 3-23, only the column vapor line and one condenser blow-off line is shown. All others are reflux lines.

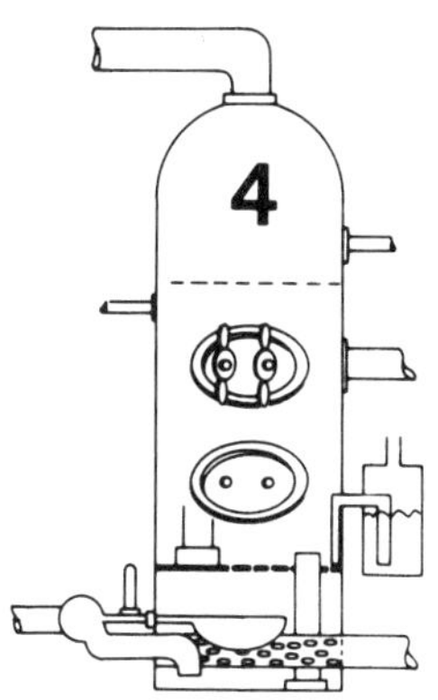

Shell And Tube Condensers

Almost everyone is familiar with the copper coil in a bucket full of water used to condense alcohol vapor to liquid. On a small scale such a device works fine; but past a certain point, such a device will not work at all. Imagine, if you will, a column in a barn or basement cranking out 20 to 40 gallons an hour. Past a certain size, or diameter, the copper coil will not transmit enough heat to the surrounding water to lower the temperature of the alcohol vapor to cool it to liquid.

You could add a number of copper condensers and water buckets to a column to get the job done, but you would certainly create a major expense and a general nuisance. You would be running all over the place checking your coils to make sure they weren't getting clogged up, tripping over the buckets, and occasionally entangling yourself in the copper tubing. *Hot* copper tubing. Imagine being the size of a mouse and trying to get through a woman's hair curlers and dodging a hot curling iron at the same time.

You won't ever see a mess like this in a commercial or industrial plant. They know better. In the distillery industry, the condenser, or condensers, on a column are usually shell-and-tube condensers. You can buy them commercially or you can junkyard one together.

Most people look at a shell and tube condenser in a distillery and automatically make the wrong assumption that the alcohol vapor goes through the tubes. It can, but not usually. Just imagine an airtight moonshiner's bucket that contains alcohol vapors, with

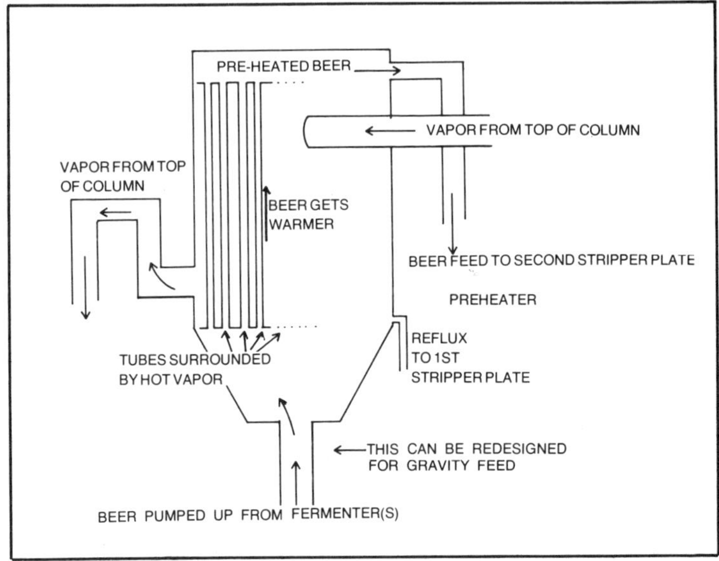

Fig. 4-1. A commercial preheater.

the cooling water circulating through the copper coil, and you will have an idea of how a shell-and-tube heat exchanger normally works. The bucket would be the shell and the copper coil would be the tube or tubes. Such a definition is technically incorrect, but what I'm trying to do here is teach a concept—not write something for another technician.

PREHEATER

Heat exchangers are normally used two or three times around stripper and rectifying columns, preheater, dephlegmator (covered in Chapter 3) and condenser. The primary difference

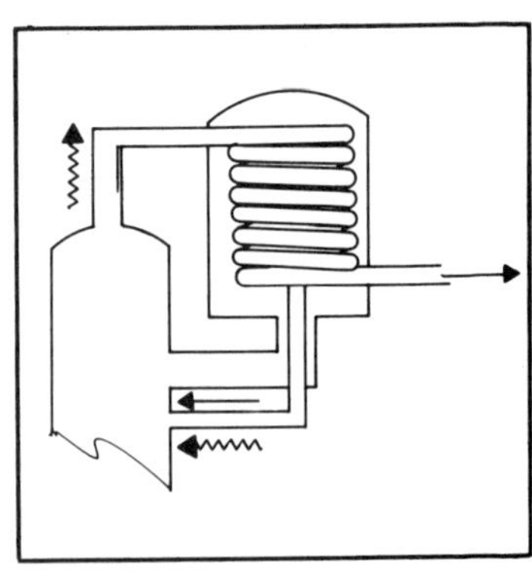
Fig. 4-2. A moonshine type of heat exchanger.

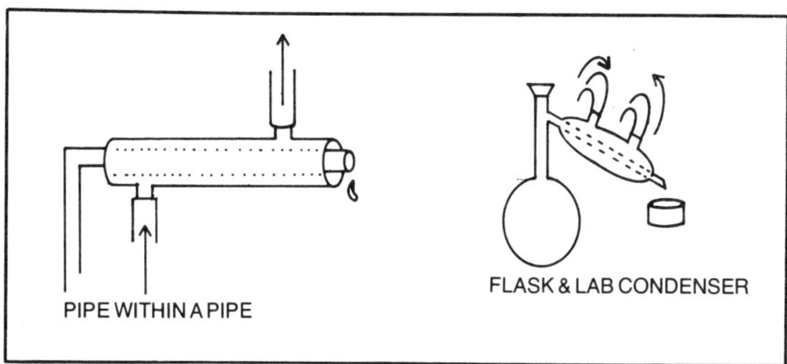

Fig. 4-3. The cooling water flow path.

between the preheater and the other two units is that the preheater does not require an outside source of cooling water. Figure 4-1 is a diagram of a commercial preheater.

The function of the preheater is to warm the incoming mash being pumped into the stripper column by scavenging heat from the hot vapor coming off the top of the column. A moonshine type of heat exchanger can be constructed as shown in Fig. 4-2.

The vapor line from the top of the column—in a two-column setup, the stripper column—is coiled inside a bucket or drum. The coil exits through a hole cut in the side of the drum and travels to a dephlegmator or condenser. Vapor travel is represented by the squiggly arrows in Fig. 4-2. The mash from the mash tub passes through the bucket until it enters the column onto the stripper plates. The double-lined arrow in Fig. 4-2 represents a reflux line from the bottom of the coil to one of the stripper plates. You can live without it, but you are taking one heck of a chance. If the mash being pumped in cools the alcohol vapor to a liquid before it leaves the preheater, creates back pressure, and stops up the coil, the resulting pyrotechnics will cost you all your mash and all your alcohol almost simultaneously. If you happen to be smoking at the time, I hope your life insurance is paid up. The alcohol vapors will create the type of fireball you normally see only in science fiction movies. White hot! Alcohol won't explode in the open air, but it will burn white hot and lightning fast. An ounce of prevention is worth a pound of cure, etc. Commercial heat exchangers will often have provisions for reflux built into them at the factory.

HEAT EXCHANGER

Let's take a look at the basics of construction of a shell-and-tube heat exchanger and see how they work. It's always easier to

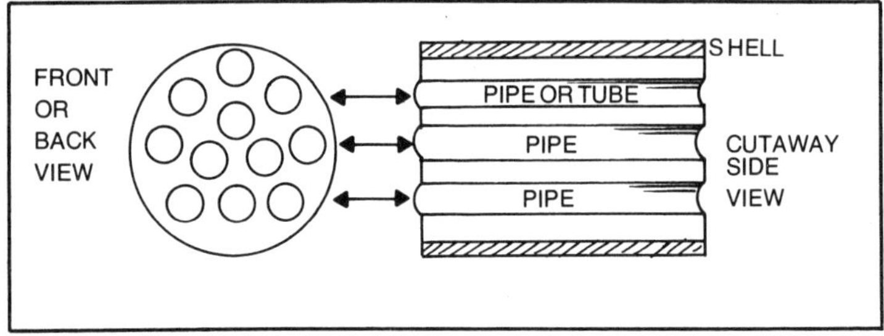

Fig. 4-4. The basis for the shell and tube condenser.

construct or duplicate an item if you know the principles involved. The simplest type of heat exchanger is a pipe within a pipe. If you run cooling water through a pipe and alcohol through the outer pipe—or vice versa—the alcohol vapor will eventually condense to a liquid. This same pipe-within-a-pipe method of doing things is the standard condenser used in chemistry labs. Only in the lab the pipes are glass.

The arrows in Fig. 4-3 represent the cooling water flow path. Simply multiply the number of pipes within a pipe (or shell) and you have the basis for the shell and tube condensor (Fig. 4-4).

Unfortunately, there is a little more to it than just stuffing a group of pipes inside another pipe. The cooling water and alcohol vapor and liquid have to have places to enter and exit. The pipes have to be braced. And the shell-and-tube condenser itself has to be mounted in such a fashion that the pressure of the mounting bolts doesn't twist and crack the shell. A vapor relief valve is installed outside the condenser as a safety feature.

Start with the construction of the vapor relief valve because it is the most obvious part of the condenser. It's out in the open where you can see it and it can be used as an indicator of the need to slow down the rate of alcohol vapors or speed up the amount of cooling water going into the condenser.

The liquid alcohol exit-pipe is where you install the vapor relief valve. It doesn't hurt to stick safety valves and pressure gauges anywhere you can, but this one is a must. See Fig. 4-5.

The pressure relief valve is simply an empty cylinder with three openings. Most of the openings are usually mounted to the adjacent pipes with pipe flanges; but it's not the only way to do things. The flanges simply make the unit more rigid. The way it works is as follows.

Alcohol vapors enter the top of the condenser (squiggly arrow in Fig. 4-5) and are condensed to alcohol (straight arrow in Fig.

4-5). If the alcohol fails to condense, it will remain as steam and create pressure. If the pressure is not relieved, an explosion might result. The pressure is relieved by allowing the steam to escape through a hole in the top of the valve, up the exhaust line (arrow with dotted line in Fig. 4-5) and outside the building into the open air. You don't want the venting of alcohol steam to take place inside a building. Again, there is the danger of fire.

If you see wisps of steam leaving the exhaust pipe, you don't have to shut your operation down. It simply means that some of the alcohol is escaping because your cooling water is too hot. The normal way to cure this is to simply speed up the rate at which you pump water into the condenser. The faster the water enters and exits the condenser the faster heat is carried away and the cooler it gets.

Which brings us to another point. Heat can only be transferred from the warmer device, item, body or area to a cooler one. If you even figure out how to reverse the process let me know. You will have just invented a perpetual motion machine.

Constructing and installing the shell-and-tube condenser can be done one of two ways. First, you could simply buy one. Where this gets tricky is in selecting the shell and head types. Each shell and tube condenser comes in three sections (Fig. 4-6).

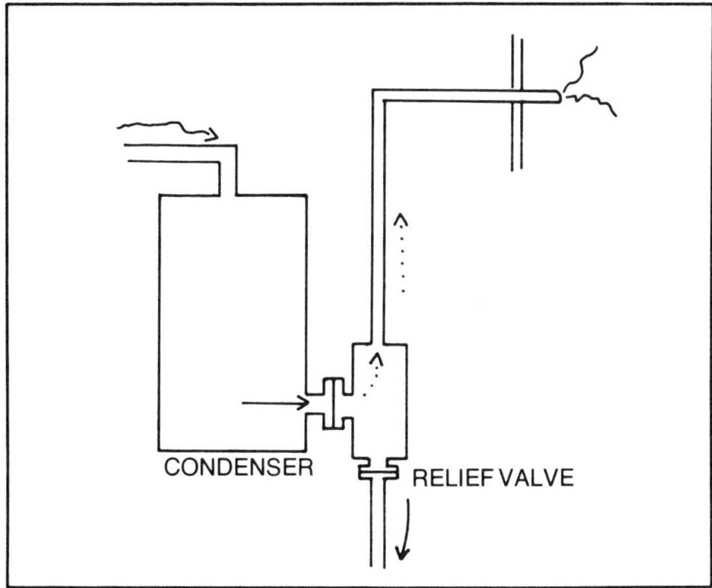

Fig. 4-5. The vapor relief valve.

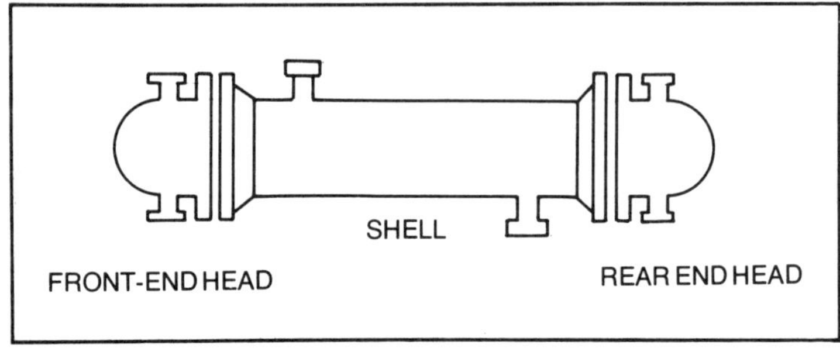

Fig. 4-6. A one-pass shell.

The reason the two heads look the same is because—according to the Tubular Exchanger Manufacturers Association—a "B" front-end stationary head is identical to an "M" fixed tubesheet rear end. Three rear-end heads are identical to front-end heads, but five rear-end heads aren't. The shell illustrated in Fig. 4-6 is a one-pass shell, type "E".

If you ordered the one illustrated, you would have to specify inside diameter and length along with the letter designations for heads and shells. An order for one with a 12-inch inside diameter and tubes 6 feet long would look like this:

SIZE 12 - 72 TYPE BEM

The 12 is for the 12-inch inside diameter, the 72 is 6 feet converted to inches, B is the front head, E is the shell type, and M is the rear head. So much for all the fancy stuff.

JUNKYARDING

Let's try junkyarding one together. Obviously, the quality of workmanship and materials is going to suffer somewhat, but what the heck, it's inexpensive. The tube bundle can be assembled two ways. One method is straight through, pipes within a pipe as shown in Fig. 4-3, and the other is a U-tube or U-tube bundle (Fig. 4-7).

The brace holding the U-tube (as shown in Fig. 4-7) constitutes a U-tube bundle rear head. "U" would be substituted for "M" in BEM. Such an arrangement is often used with outdoor storage tanks for heavy fuel oils and molasses, only for heating. Steam is piped in without the necessity of leaving more than one opening in the tank for steam entrance and exit (Fig. 4-8).

The primary disadvantage to the U-tube bundle is that it is a bear to clean. Imagine trying to run a bristle brush down the barrel of a rifle looped like that and you will understand what I mean. Stick with the pipes within a pipe method of doing things.

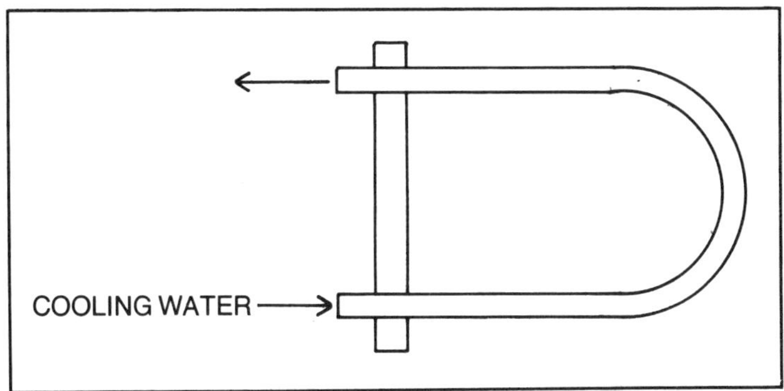

Fig. 4-7. Brace holding a U-tube.

For the shell, you can use an old water heater, pipe from a culvert or a 55-gallon drum. Any type of heat-resistant pipe will do. Keep the pipes and the cylinders upright so that the hot vapor will remain in the top of the condenser and the condensed liquid will fall to the bottom.

Again, the cooling water goes through the pipes, the vapor condenses to liquid on the outside of the pipes and the inside of the shell, falls to a sump in the bottom or back head and flows out into the pressure relief valve. Metal plates with holes for the pipes—they look like thin manhole covers—should be bolted or welded at each end of the cylinder or shell (Fig. 4-9).

Bolting is better in the long run because (if you thread the pipes and the holes they fit in instead of welding) the entire apparatus can be disassembled for cleaning and repair of each individual pipe. If one pipe becomes clogged with scale or corroded so badly that it leaks, it can simply be unscrewed and replaced.

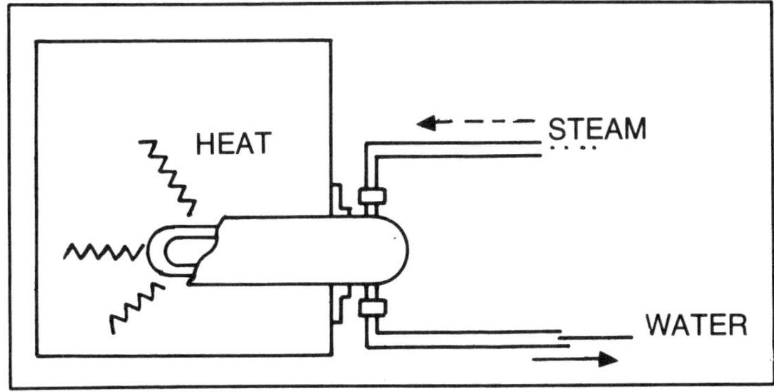

Fig. 4-8. Steam is piped in only one entrance.

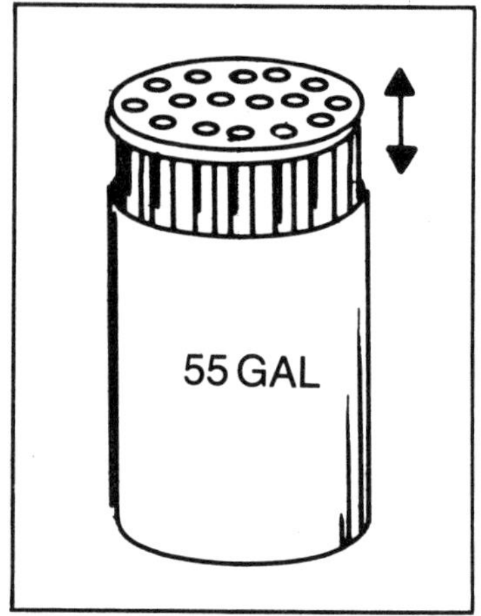

Fig. 4-9. Metal plates with holes are bolted or welded at each end.

The top and bottom (or front and rear) heads merely allow for one large hole—instead of a number of little ones—to pump the water in and out. Your completed junkyard shell-and-tube condenser will look something like Fig. 4-10.

The straight arrows in Fig. 4-10 represent the water being pumped into the bottom, up through the pipes and out the top. If

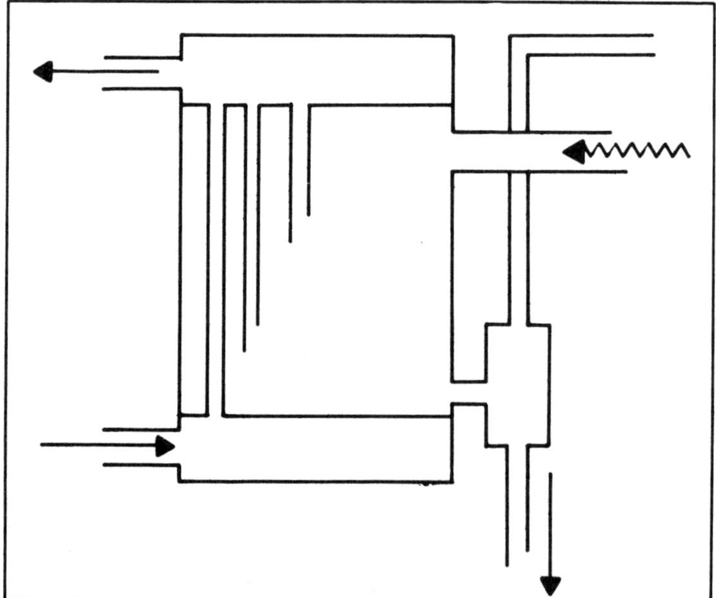

Fig. 4-10. A junkyard shell-and-tube condenser.

you don't have a pump, simply reverse the flow and let gravity do the rest. With a pump, the water should be pumped in the bottom because you do not want to be carrying heat toward the cooler part of the condenser—the bottom—if you can avoid it.

The squiggly line in Fig. 4-10 represents alcohol vapor from the column and the arrow pointing straight down represents liquid alcohol flowing out of the pressure relief valve. For the top and bottom of the condenser—the heads—almost anything you can keep from leaking water will work. The bottoms sawed off a few additional 55-gallon drums should do nicely.

In an extremely dry area, you can simply throw the middle section away, run the hot vapor down through the pipes, and cool it with a fan. A large fan. In some of the really windy states, you could probably make an extra long tube bundle and throw the fan away.

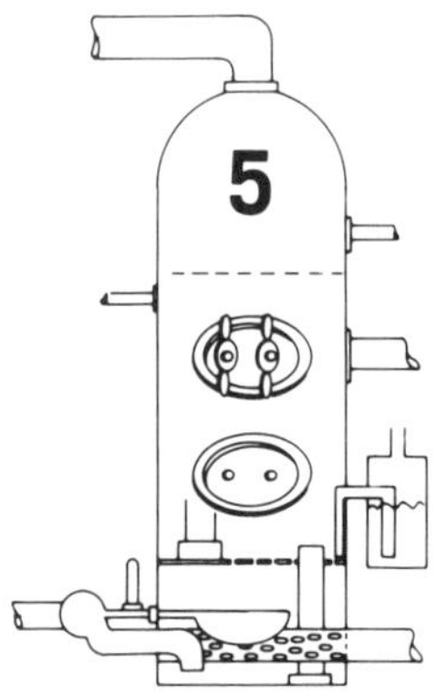

Building A Junkyard Steam Boiler

In any type of distilling operation, you will have to have some way of generating heat. The most obvious way is to simply build an open fire under whatever it is you're trying to cook. Sometimes this can cause problems.

The biggest drawback to the open fire method is the amount of time required to get the fermented mash hot enough to boil. If you have 200 gallons of mash in a 250-gallon tank, every single gallon has to absorb heat before anything at all can be distilled. This can take up to four hours. On a stripper column, the process begins as soon as the mash being pumped in and steam coming up from the steam line collide.

Another drawback to the open fire method is the tremendous amount of heat that escapes into the atmosphere. You can imagine how many btu's get carried away into the surrounding air during a four-hour warm-up period.

Before you can generate steam for utilization in your distillery operation, you have to have a steam boiler. Start with a simple one.

A word of caution first.

When you're dealing with steam, you are dealing with one of the most explosive forces known to man. It was water turning to steam that blew the lid off Mount St. Helens in 1980. That explosion was equal to 2500 of the atomic bombs the Enola Gay dropped on Hiroshima on August 6, 1945. Any explosion you create with a steam boiler won't be a tame one. You would be in less danger throwing a hand grenade into a barrel of water next to you than if you had a steam boiler go off in your face.

Be careful.

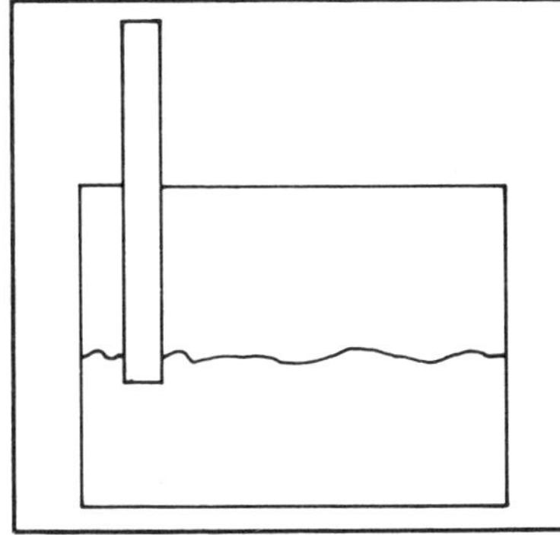

Fig. 5-1. Cut a hole for the pipe and weld it in place.

A SIMPLE DESIGN

The simplest type of steam boiler can be made out of a 55-gallon drum. Containers do come in larger sizes such as 250-gallon propane tanks. The most common mistake in building a steam generator is to build it too small. A 55-gallon steam generator simply will not produce enough steam to run a 12-inch diameter column. If 55-gallon drums are all that's available, you can hook them up in series to make 110-, 165- and 220-gallon steam generators.

First you must have a way to fill the drum. This is done by simply cutting a hole at one end of the drum lid and welding a pipe in place. The pipe should terminate about 6 inches from the bottom of the 55-gallon drum. See Fig. 5-1.

The generator can be "charged" simply by sticking a garden hose down the pipe and turning a faucet. If the steam pressure becomes too great, water will blow back out the pipe and flip the hose out. If the water level falls below the bottom of the pipe, steam will escape and alert the operator to turn the hose on and fill the boiler.

The next step is to simply screw a pipe in the bunghole for the steam to escape. You will probably have to use elbows, couplings,

Fig. 5-2. Using elbows, couplings, nipples and flanges.

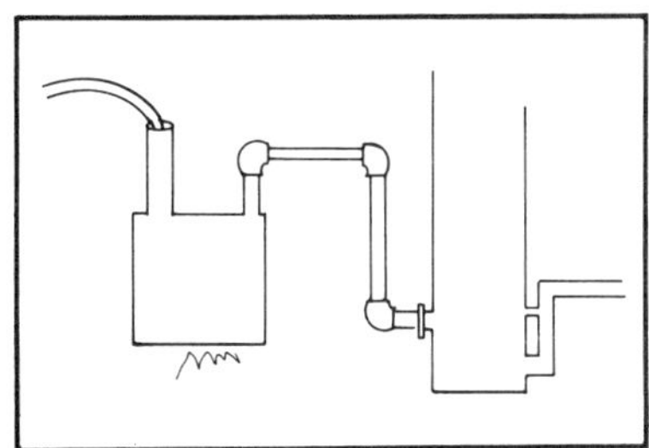

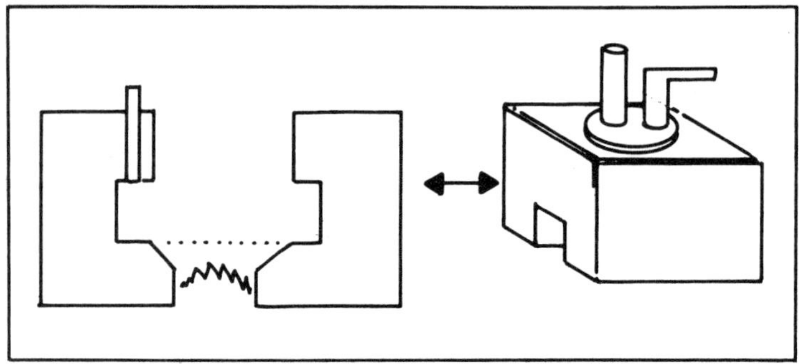

Fig. 5-3. A brick furnace.

nipples, and flanges to get the steam from the bunghole to the bottom of your column. It might look something like Fig. 5-2.

You will probably want to install a pressure gauge on the lid of the drum to give you a more accurate idea of what is going on. You need less than 5 pounds of steam pressure. Watching a gauge is preferable to being periodically drenched by hot water or steam. It also saves fuel. Wild fluctuations in heat and steam pressure cause the steam generator to consume more combustibles. Corn cobs are a great source of fuel. One bushel of corn cobs will normally provide enough heat to process 1 bushel of corn.

If you want to get fancy, you can install a drain valve in the bottom of the drum. Normal boilers develop corrosion and scale problems and need to be cleaned and flushed periodically. I suppose you could get emotionally attached to a 55-gallon drum, but it makes more sense to just throw it away and get another one. You'll know when it's time; pinhole leaks will appear in the drum and steam will escape. When that happens, shut it down and replace it. Scale won't do any harm unless it clogs the pipes.

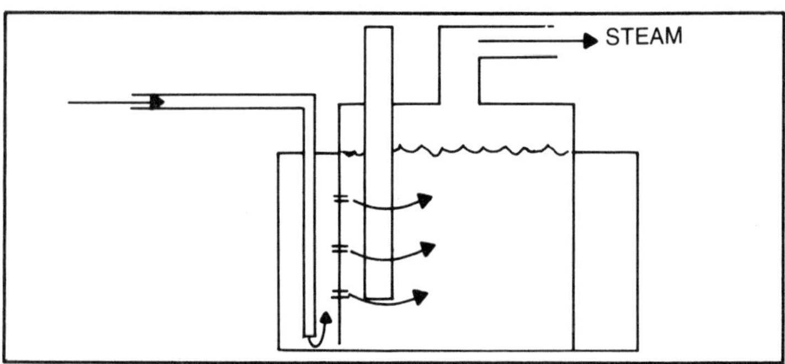

Fig. 5-4. Preheat the water.

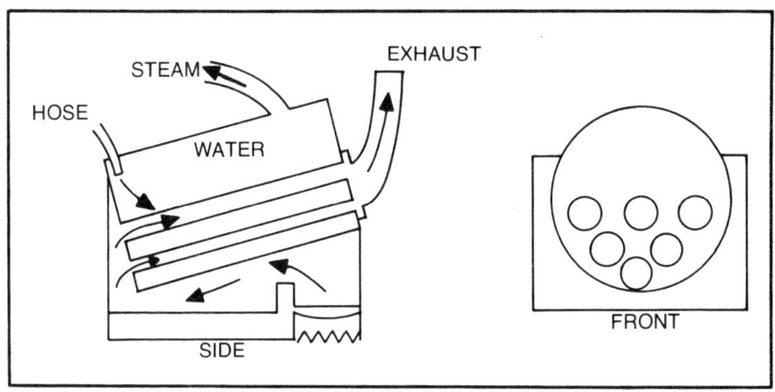

Fig. 5-5. The layout for a tube boiler.

If you intend to pursue the production of alcohol on a regular basis, you might want to brick up a furnace to set the drum in. The furnace keeps much of the heat from escaping into the air. Furnace brick work will look something like Fig. 5-3 before the drum is placed in it.

The dotted line in Fig. 5-3 represents a grate or other support for the drum. The ring or annular space at the bottom allows the heat to circle the drum before it escapes into the atmosphere. With a loose fit, the smokestack is not necessary. Of course, a loose fit means wasted energy.

The steam generator just described is going to have to be placed extremely close to your column. The reason is that steam in contact with water leaves at about the same temperature of the water: 212°F. As soon as the steam does any work—which includes sliding along a steam pipe—it loses heat and a portion of it condenses. Steam that is the same temperature as the water it leaves is called *saturated steam.* If you're trying for the 550°F temperature used in the bottom of a commercial column, you're not going to make it with saturated steam.

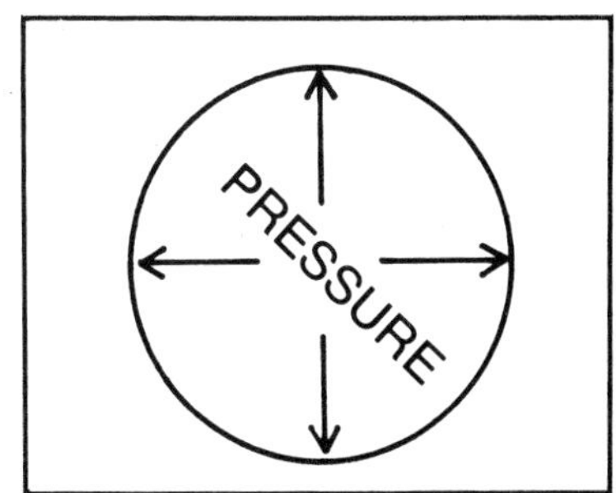

Fig. 5-6. A sphere is ideal for pressure control.

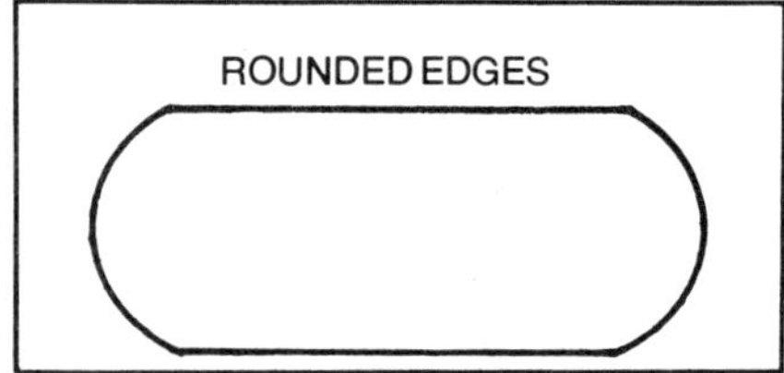

Fig. 5-7. A typical boiler shape.

What you want is *superheated steam.* The way you create superheated steam is to heat the saturated steam in a vessel separate from the boiler. Superheating the steam helps avoid condensation and increases the efficiency of your operation considerably. Another method of increasing the efficiency is to preheat the boiler water using existing sources of heat.

A SOPHISTICATED JUNKYARD BOILER

I will outline a somewhat more sophisticated junkyard boiler now that you see how easy a pot boiler is to build. First, pre-heat the water to be boiled. Simply take the drum you are using for a boiler and plop it into a larger drum. All you are doing is insulating the main boiler; heat is transferred from the main boiler to the water in the larger drum which in turn feeds warm—as opposed to cold—water into the boiler. See Fig. 5-4.

Such preheating and insulation will quite often cut the fuel consumption of the boiler by as much as a third. Another method of increasing the efficiency of the boiler is to use tubes that are similar to shell and tube condensers. Whether you use water or heat in the tubes, be sure to incline them slightly so that the passage of water or steam is up a slight grade. A completely horizontal tube is dangerous because material can "collect" and a vertical tube wastes energy.

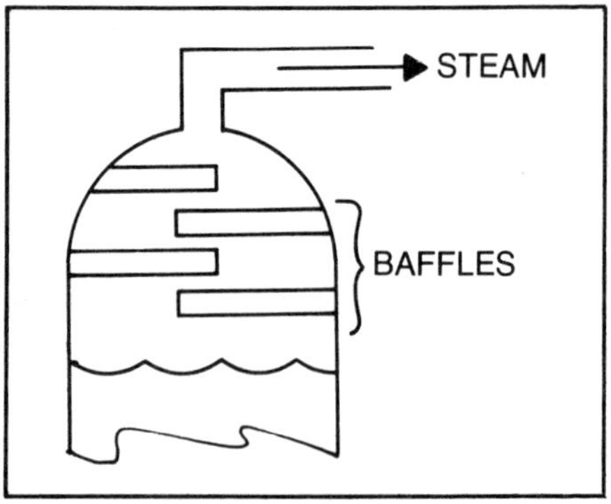

Fig. 5-8. Superheating steam.

If you think a tube boiler will amuse you, the layout looks something like Fig. 5-5. The tubes simply increase the heating surface in contact with the water.

Before I get to the superheater, a few words of caution are in order. If you are going to weld your own boiler together instead of using a drum, try to avoid flat surfaces. They are easily deformed. A sphere is ideal for pressure because everything equalizes. See Fig. 5-6.

Obviously, a true sphere—though ideal—is not practical because of mounting, space and other considerations. A typical boiler will be something of a compromise. See Fig. 5-7. You might see some commercial boilers that appear square from the outside, but they aren't built that way internally.

You can use a thermometer as a pressure gauge. If the temperature of the saturated steam is 240°F, then you are operating at 10 pounds per square inch above atmospheric pressure. If the temperature of a junkyard assembly goes over 270°F, then you are operating pretty close to the wire—30 pounds per square inch over atmospheric. Remember, we are talking about odds and ends being used to get the job done. Some commercial boilers are in no danger at 1500 pounds per square inch.

Do not use silver solder on any of the fittings. It won't hold. Copper can be used for the tubes if you can stand the expense. It transmits heat seven times more efficiently than iron. If you do use copper, be sure to anneal it when you are working. Heat it to a dull red in subdued light and allow it to cool. Copper is a hard metal to thread. Be sure to use clean taps and dies. Use turpentine as a lubricant. You can minimize corrosion by filling the boiler with clean boiled water when not in use.

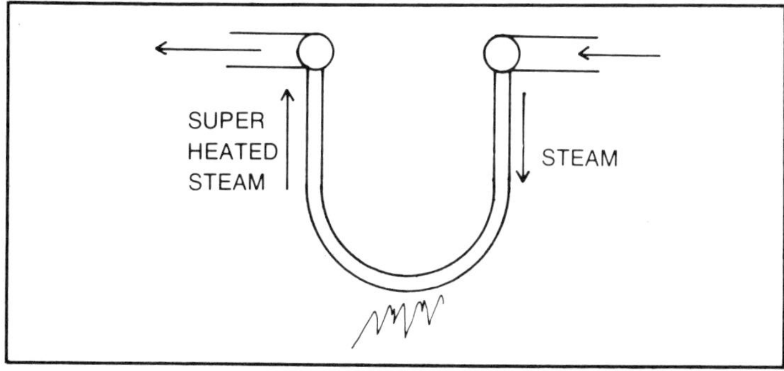

Fig. 5-9. A "U" shaped superheating pipe.

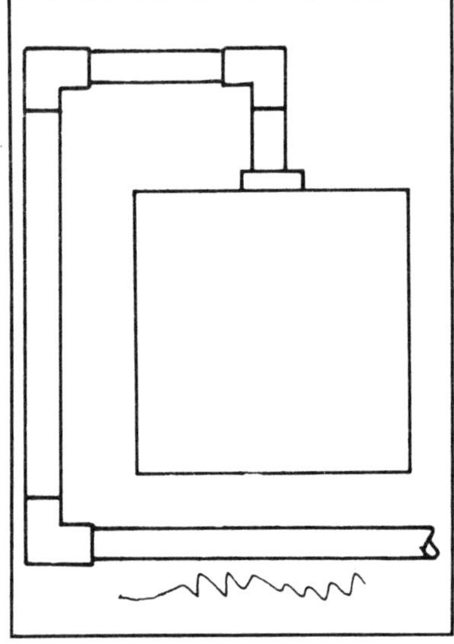

Fig. 5-10. The superheater pipe can be run straight through the fire.

THE SUPERHEATER

And now for the superheater. There are two items to keep in mind:

—you want to superheat saturated steam. That means that you now want to heat steam, not water.

—you don't want to build a separate fire.

Here's how it's done. First, you must put baffles in the top of the boiler. You don't want *priming* (carrying over the water with steam). The carry-over of water is fatal to the functioning of a superheater. See Fig. 5-8.

Water will condense on the baffles and return to the boiler. If the boiler "burps" the water will simply splash against the baffles and fall back.

The saturated steam is then run into a pipe that is 50 percent greater in cross-sectional area (not diameter) than the saturated steam pipe. Frictional losses do occur.

The superheating pipe can take one of three basic shapes. It can bend down near the fire in the shape of a "U" (Fig. 5-9). This is the type most commonly used in the type of boiler in Fig. 5-5.

The superheater pipe can also be run straight through the fire (Fig. 5-10). Or it can be run through the fire in the form of a coil. This is known as a *flash boiler.* Just be sure that you make provisions for reading the temperatures and pressures along the way.

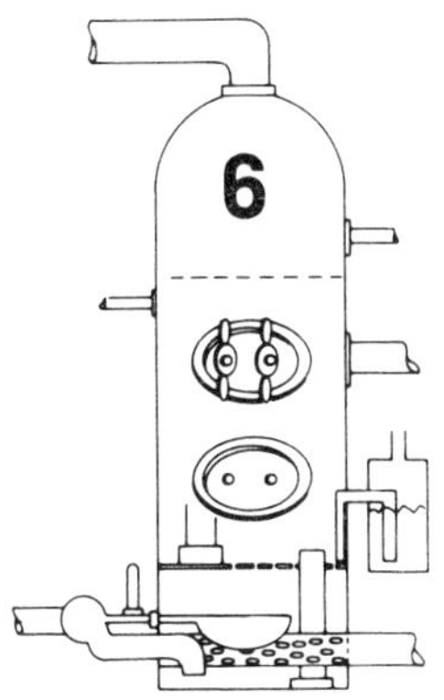

Protecting Your Still, Yourself and Bystanders

In any distillery operation, what you are dealing with is steam. Most backyard operators don't give a second thought to safety precautions, but you should. There aren't many more painful ways to go than to be scalded to death in a boiler explosion.

One item I haven't seen mentioned anywhere is the problem of force-per-unit area. Several years ago, the floors in the 707 aircraft were constructed of two thin sheets of metal with a honeycomb material sandwiched in between. It passed every engineering test imaginable.

And it failed the first time it was tried.

What happened was that the high heels of the airline stewardesses punched right through the metal. This is caused by a little quirk of physics known as force-per-unit area. A 110-pound woman with slippers on distributes 110 pounds of force over about a one-half square foot area (or 220 pounds per foot). The same girl walking down the aisle of an aircraft puts her weight on a stiletto heel of less than ¼ square inch unit area. If you divide that ¼ square inch into a square foot (144 square inches), that ¼ square inch has a force equal to 1/144 ÷ ¼ x 110 or over 63,000 pounds per square foot.

PRESSURE RELIEF VALVE

It is fairly easy to put a 6-inch column on top of a 500-gallon cooker, stoke up the fire underneath, and create a steam-powered version of an incoming artillery round—complete with flying shrapnel. If you install a pressure relief valve, the only thing that

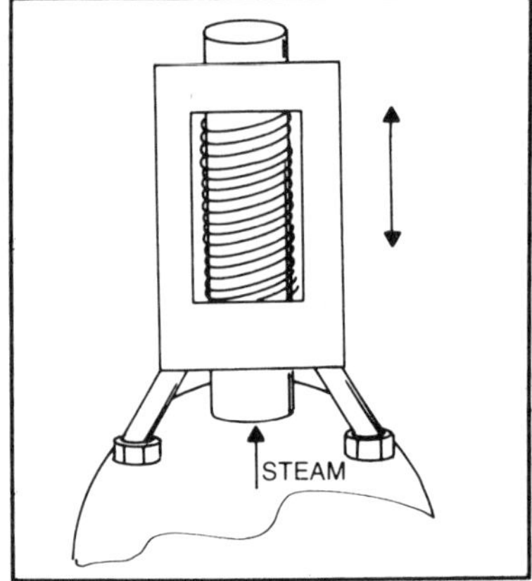

Fig. 6-1. A springloaded pressure release.

will happen is that a lot of alcohol and steam vapor will escape periodically into the atmosphere. This is expensive but safe.

A common type of pressure release is the springloaded type. At a certain pressure, the spring compresses and the plug on the top of your column or cooker simply rises enough to let some steam escape (Fig. 6-1).

VACUUM INTERRUPTER

Another item that seems to be often overlooked by the barnyard builder of alcohol columns is the *vacuum interrupter*. If cold air or water enters the column while it is in full operation, a vacuum is created and the still quite often cracks. The vacuum interrupter is simply a round plug of metal attached to an arm with a counterbalance. When a vacuum is created inside the column, the plug drops through the top of the column and equalizes the

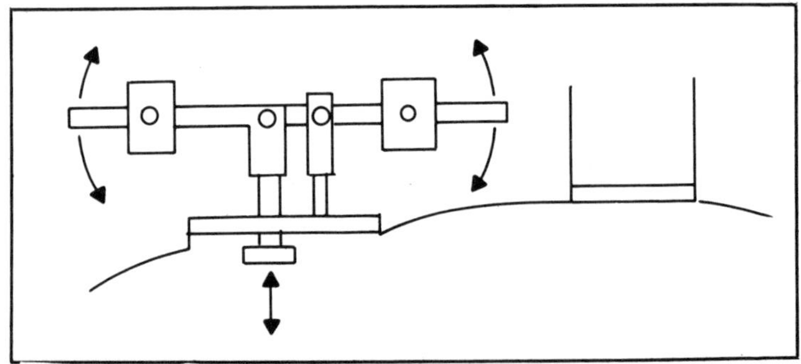
Fig. 6-2. Pressure relief valves.

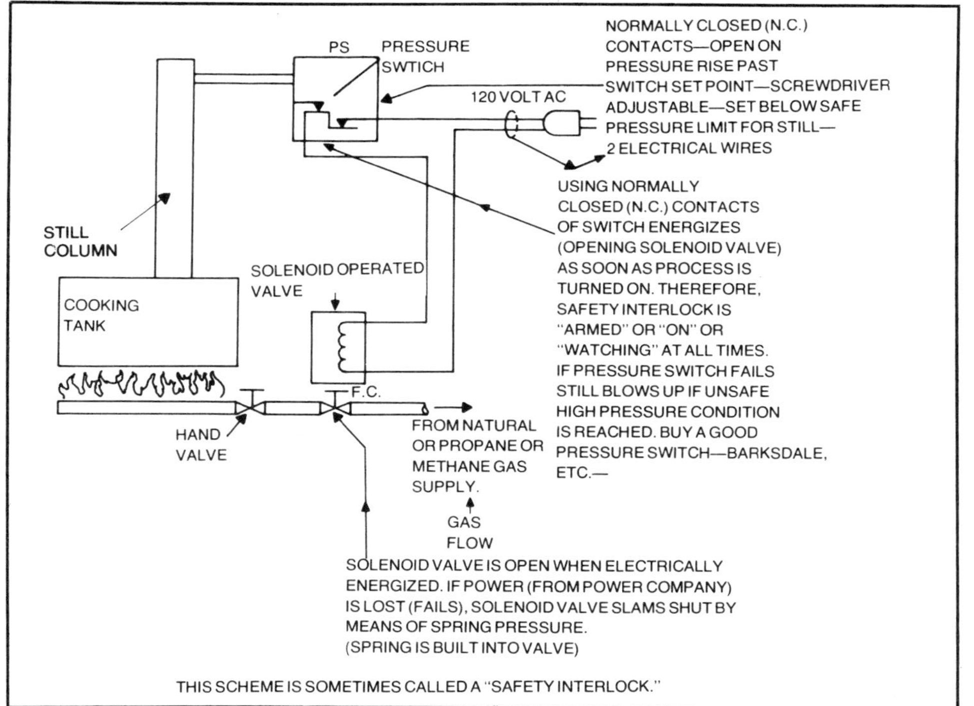

Fig. 6-3. Safety pressure override for a column still and cooker.

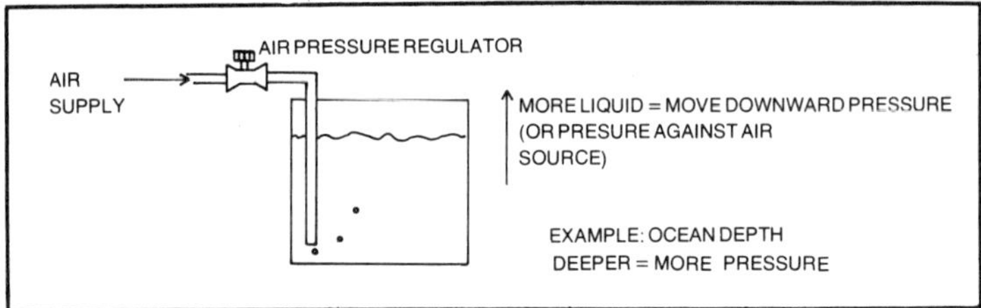

Fig. 6-4. The bubbler.

atmospheric pressure both inside and outside the column. The counterbalance pulls the plug back into the hole the instant the pressure is equalized.

In short, use a pressure relief valve to keep the still from exploding outward and a vacuum interrupter to keep the still from cracking inward. See Fig. 6-2.

The primary problem with pressure relief valves and vacuum interrupters is that they only work "after the fact." When they go off, it's like trying to lock the barn door after the horse is out. In this case, you are merely keeping the horse (or alcohol and steam) from wrecking the barn (or alcohol still) on the way out.

INSTRUMENTATION

The way to prevent such problems is with instrumentation. In the long run, it is much easier to watch a few dials and gauges in order to prevent a problem before it gets out of hand. In many cases, instrumentation will also prevent financial loss by alerting you to leaks. You can control the situation before anything serious goes wrong.

The following are the four basic quantities measured with instruments:

—Pressure.
—Flow.
—Liquid level.

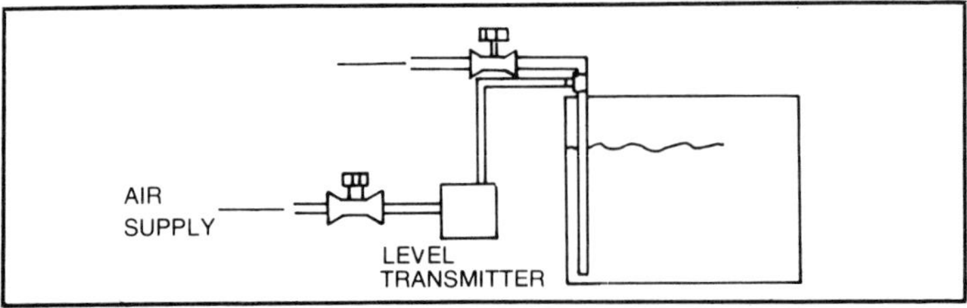

Fig. 6-5. A practical bubbler system.

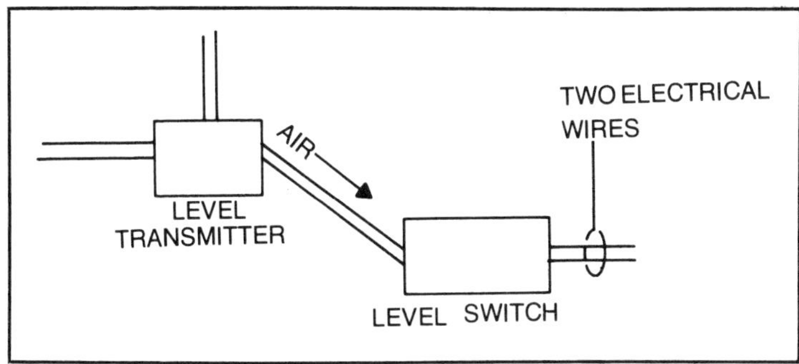

Fig. 6-6. A level transmitter converts liquid pressure to pneumatic pressure.

There are many other items that can be measured—such as pH—with instrumentation. However, the preceding four are the most important.

A really crude system is simply a stick with inches or feet marked on it. At the turn of the century, such a device was used as a "gas gauge" for automobiles. In the terminology of the trade, such a device is known as a *visual indicator.* That is, you can see what is going on.

The other two types of instrumentation are *blind* and *control*. Control instrumentation will not only indicate what is going on, it controls the problem before it gets out of hand. A classic example

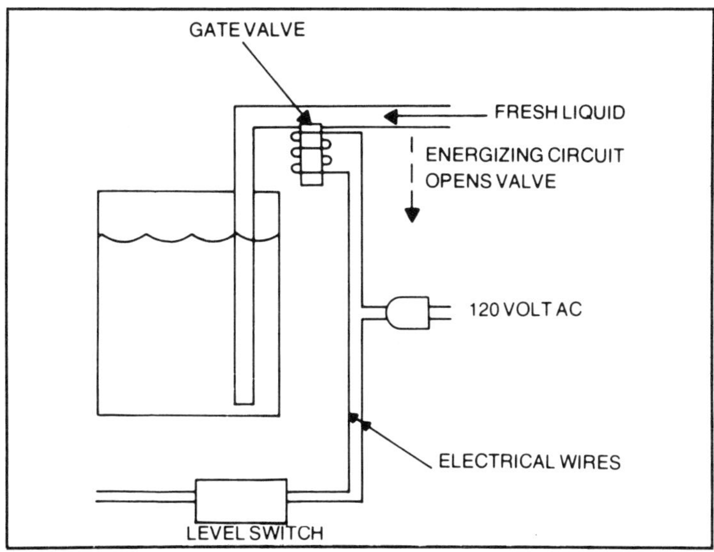

Fig. 6-7. This system is ideal for butanol fermenting operations.

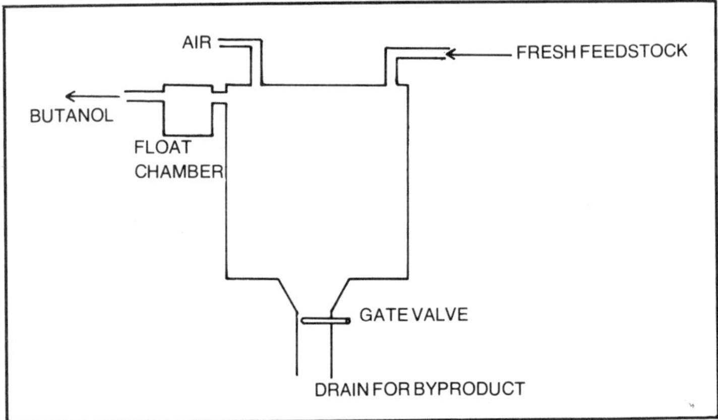

Fig. 6-8. Don't forget to occasionally clean the byproduct.

of this type of instrumentation is a pressure switch. Such a system works as shown in Fig. 6-3. If you see a black box on a commercial column, it is usually control instrumentation.

An example of blind instrumentation is the float and valve arrangement inside the float chamber of a carburetor or a toilet tank. In the trade, the float is known as a *primary transducer* or *energy changer*. The energy in the rising fluid is changed into a mechanical movement in order to close a valve and shut off the further flow of fluid.

If you mounted a sight glass on the toilet tank, that would be known as a *level gauge* or *level indicator*. If you are dealing with something other than clear water or other liquid—such as a slurry so full of particles it will clog up a sight glass—use a *bubbler*.

The bubbler is simply a supply of compressed air pumped into the bottom of your process tank, such as a 55-gallon drum, that is regulated by a springloaded and adjustable air pressure regulator. The amount of liquid in the tank creates a certain amount of back pressure in the compressed air line. When the liquid falls to a certain level, the pressure of the compressed air overcomes the water pressure and bubbles escape. See Fig. 6-4.

This might seem a little strange because you could tell what the liquid level was without looking at (or needing) a compressed air line to make bubbles. To make the system practical, add a level transmitter with its own air supply and a pipe and t-fitting connected downstream from the pressure regulator (Fig. 6-5).

The level transmitter simply converts the liquid pressure to pneumatic pressure. This in turn pressurizes a level switch

Fig. 6-9. Bill McDonough teaches classes on instrumentation.

(actually a pressure switch). See Fig. 6-6. The air coming from the level transmitter pushes against a diaphragm in the level switch. This closes an electrical circuit. A level switch will normally have a screw adjustment that allows you to set it to open or close whenever you want.

When the electrical circuit closes and the current flows through it, a solenoid valve opens and allows fresh liquid into the tank (Fig. 6-7). Such a system is ideal for a butanol fermenting operation. All you have to do is dump in fresh water and starch from an outside source, set up a float chamber to drain off the butanol, and occasionally clean out the byproduct (Fig. 6-8).

If you want someone to come out and do it for you, don't call me. The fellow you want is Bill McDonough, Rt. 1 Box 5, Thayne, Wyoming 83127. He has been in the instrumentation business for 25 years and teaches classes in the subject for do-it-yourselfers (Fig. 6-9).

The same type of setup used for the liquid level can be used on the gate valve (mount a solenoid). It's just a question of how much time and money you want to spend in the beginning to automate, instrument and install safety devices to keep from having to spend time and money later on. As the old saw goes, an ounce of preventation is worth a pound of cure.

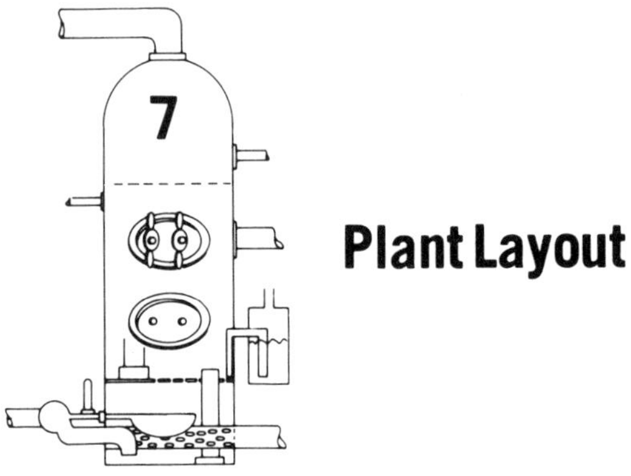

7 Plant Layout

Alcohol plants are usually divided into several separate sections or buildings. You could stuff everything under one roof, but it might cause you some inconvenience later on. For example, a Kansas or Nebraska farmer who wants it all—motor fuel, cattle feed, dry ice, and so forth—is going to need some help from his neighbors.

LOCATION

A few years back a lot of farmers put together "Co-ops" to blend their own gasohol. As far as I know, none of them were raging successes simply because of a fundamental characteristic of human behavior—too many cooks spoil the broth. It makes a whole lot more sense (to me anyway) for one farmer to have a distillery, another an evaporator room, a third a dry ice plant, and on down the line. Farms back-to-back can simply run pipelines to each other for CO_2 and wet slurry. A single farm setup would look something like Fig. 7-1.

A lot of small or medium-size cinder-block buildings allow a farmer to build onto his existing operation gradually. On-farm operations solve the major problem confronting the big oil companies in their quest for profits: transportation of the raw material to the processing plant.

How an alcohol plant is laid out usually depends on where the particular individual running it wants things for his convenience. Included in the following pages are some typical layouts for both farm and commercial installation. When you see a layout with more than two columns, those are simply to separate substances in

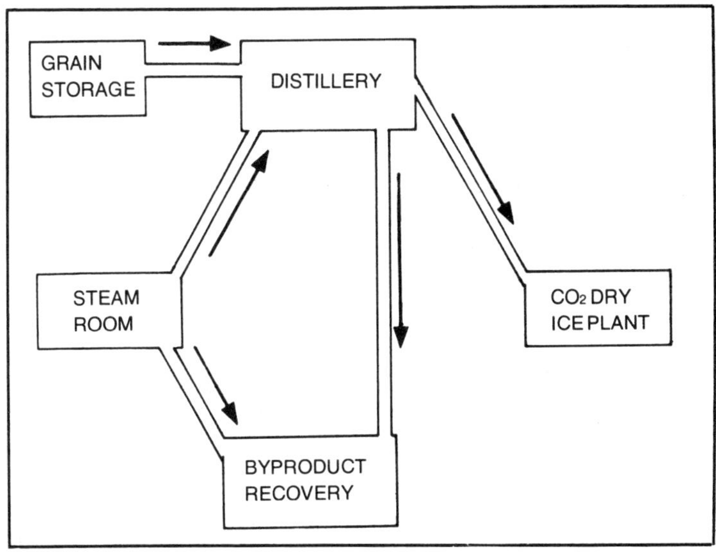

Fig. 7-1. A single farm setup.

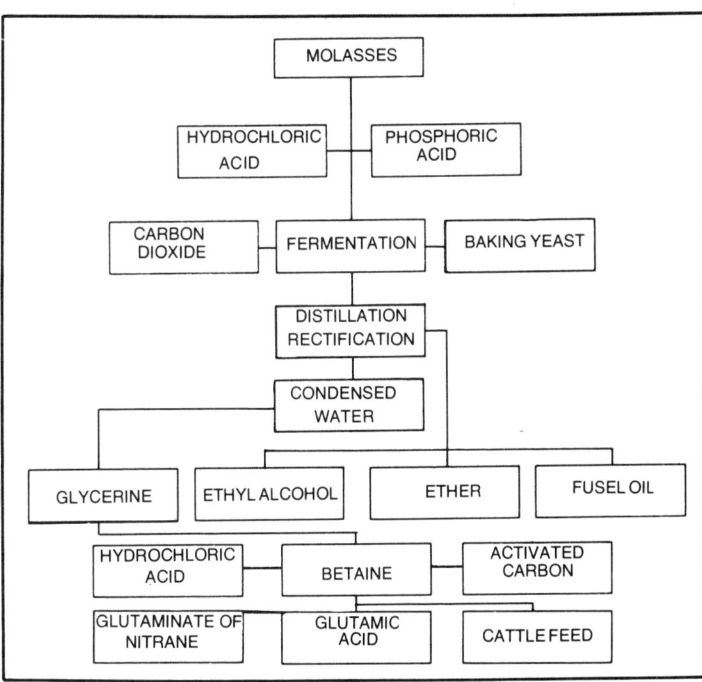

Fig. 7-2. A flowchart of industrial products available from starch or sugar products.

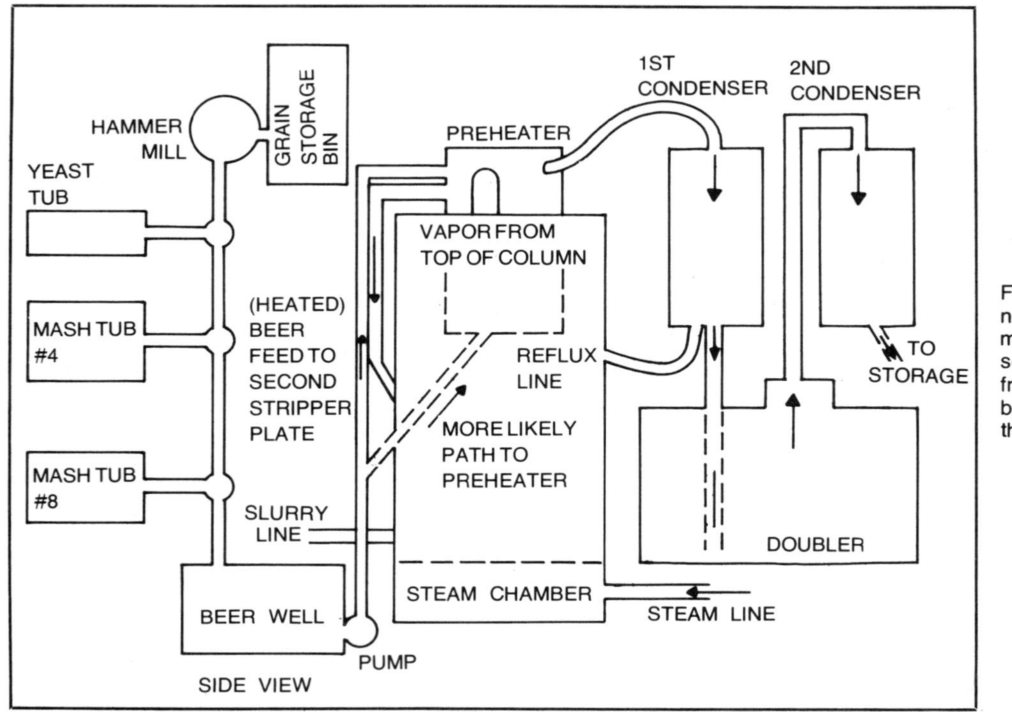

Fig. 7-3. Because fermentation normally takes three days, eight mash tubs allow you to begin a separate fermentation and run a fresh tub of fermented mash into the beer well (merely a holding tank for the column) every nine hours.

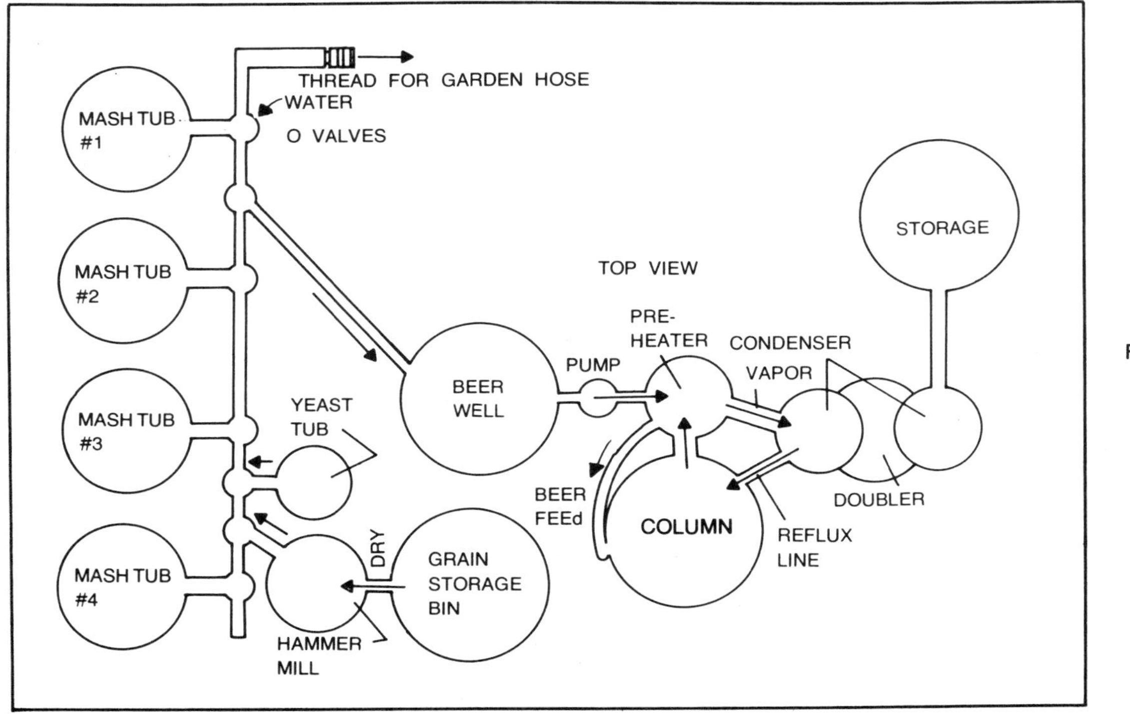

Fig. 7-4. Typical layout.

addition to alcohol and water. Such columns are normally rectifying columns for separating aldehydes and fusel oils from alcohol; the next column separates fusel oil from aldehydes. The number of columns will always be one number less than the number of items you are trying to separate. For example, if you are trying to separate four substances, you will need three columns.

SCHEMATIC

If you read this book carefully, you shouldn't have any problem following the included drawings, schematics and flowcharts (Figs. 7-2 through 7-6). If you don't understand, don't be discouraged. Just go back to the first page and try to be a little more patient.

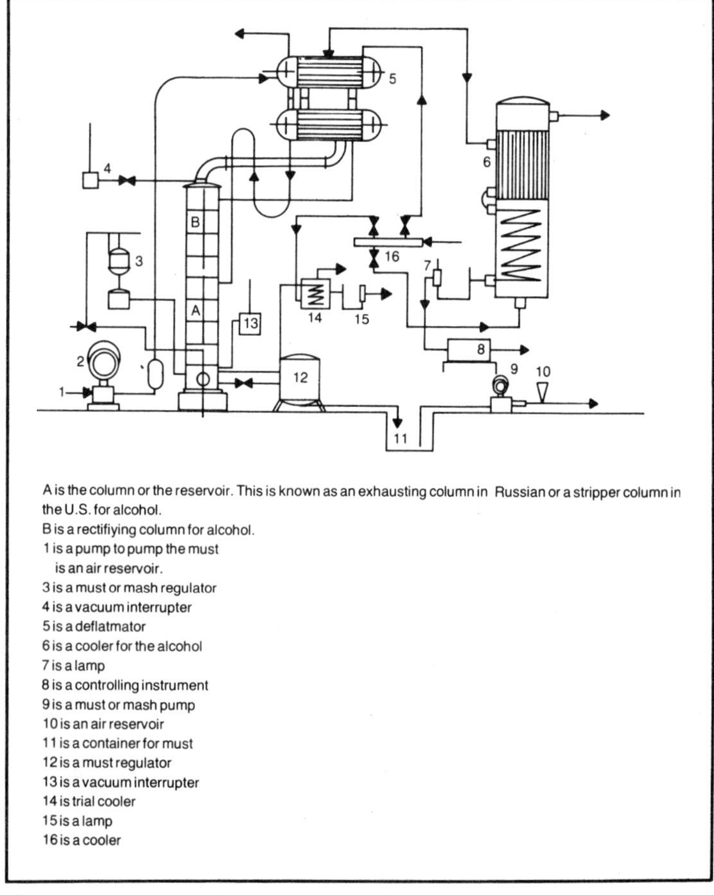

Fig. 7-5. A one-column distillation apparatus.

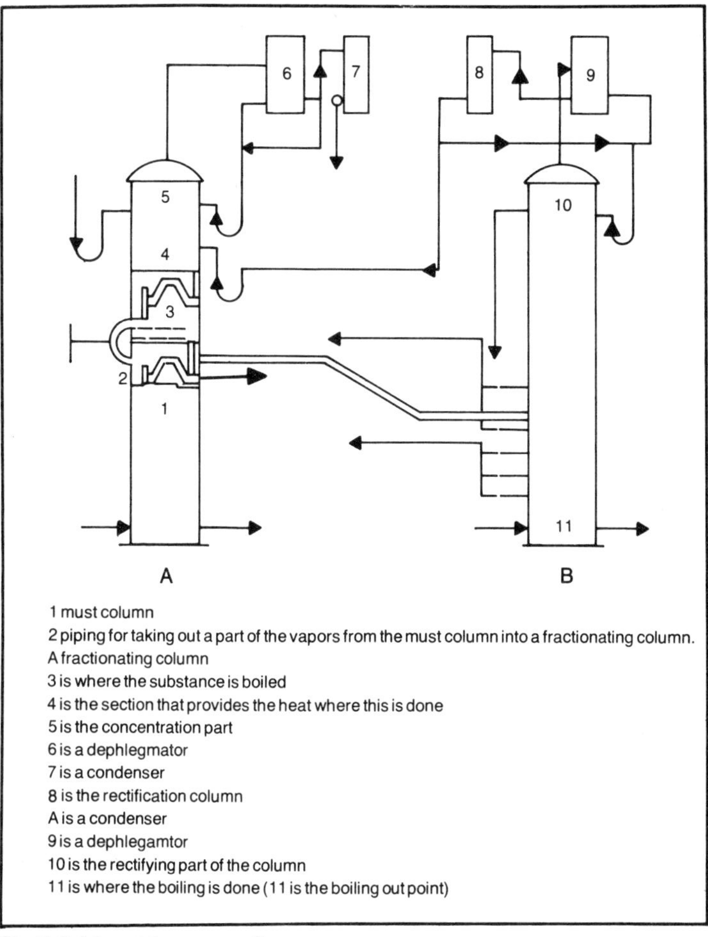

Fig. 7-6. A two-column operation.

Figure 7-2 illustrates only a few of the industrial products available from starch or sugar products. In addition to the items shown, over 80 pounds of vitamin D and over 900 pounds of vitamin B12 feeding concentrate can be obtained per ton of feedstock.

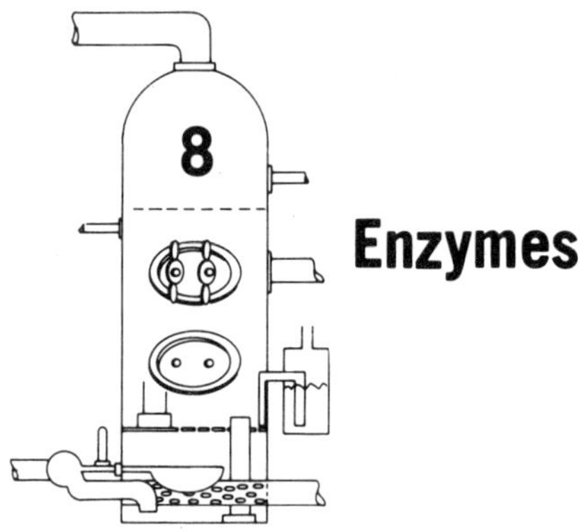

8 Enzymes

The starch-splitting enzyme used commercially is a mold. In the process of producing your own enzymes, sanitation is an absolute. Barley malt doesn't seem to be affected anywhere near as much as Aspergillus niger. When people have things go wrong in commercial enzyme production, it can almost always be traced back to bacterial contamination. Even the folks who discovered the mold now in use had to have 15 complete runs through a commercial distillery before they could get anything to come out right. They finally traced it down to bacterial contamination in their cooker. Keep in mind that the spores of some bacteria can survive five to six hours of boiling.

Amylases and diastases (meaning "I split") hydrolyze starch to dextrins. A dextrin is still several sugar molecules stuck together. In addition to enzymes, heat or acid can break dextrins down to the sugars maltose and glucose.

Starches are also known as carbohydrates. The name carbohydrates was first used because simple sugars such as glucose have molecular formulas that appear to be "hydrates" of carbon. A hydrate is merely a solid, such as carbon, that has H_2O molecules bonded to it in such a fashion as to form a separate and distinct molecule. Later research proved this theory false, but the term *carbohydrate* has remained in use anyway. It was probably too much trouble to go back and change all the books and retrain all the chemists.

Sugars, the simplest type of carbohydrate, are also called sacharides. In reading distillation literature, you will often come across the term *polysaccharide*. The term means a sugar that is capable of being divided into simpler sugars. A *monosaccharide* is a sugar that—if you break it up into smaller pieces—is no longer a sugar. A crude analogy would be polygamy, for a man having several sweeties, and monogamy—for a man having only one wife.

If you want to be completely accurate, a polysaccharide will yield more than eight monosaccharides. A *disaccharide* yields two, a *trisaccharide* three, and so on. An *oligosaccharide* (from the Greek, oligon, few) are polysaccharides that yield from two to eight monosaccharide units upon hydrolysis.

Glucose is a simple sugar consisting of dextrose, levulose or a racemic mixture. If you shine light at dextrose under controlled conditions, the light rotates clockwise. With levulose the light rotates counterclockwise. A racemic mixture is levulose and dextrose mixed together. It's all glucose no matter how you look at it.

Just in case you wondered, Aspergillus niger isn't the only mold you can use. Aspergillus oryzae can be cultured on moistened wheat bran. It is incubated under controlled conditions of temperature and air circulation. The product is then dried. From 1945 to 1946, at least one alcohol plant in the United States used this method.

Mold bran is made by surface-culture fermentation. There are several ways to grow the stuff. One method is to grow a species of Mucor or Rhizopus submerged in the grain mash itself. The starch is converted to sugar by the action of the mold enzymes and then fermented to alcohol by yeasts. This process must be carried out under the most extremely pure culture conditions. Some alcohol plants overseas still use this method.

Aspergillus oryzae can also be grown submerged in thin stillage. Sometimes you might have to get the procedure started with a small amount of malt and then add the mold culture and yeast. A similar process can be used for another strain of fungal amylase called Rhizopus delemar. Granular wheat flour is used in place of wheat bran.

ALCOHOL PRODUCTION

The process developed for the preparation of fungal amylase and its utilization for the production of alcohol is, briefly, as follows. A sterilized medium composed of thin stillage, grain, and calcium carbonate is inoculated with a liquid culture of Aspergillus niger, NRRL 337. Investigations have shown that improved enzyme preparations are obtained when calcium carbonate is eliminated from the stillage medium and the concentration of grain is increased. During incubation at 86° F, the submerged culture is aerated and agitated. The mold is propagated for 50 to 60 hours; the production of fungal amylases is then practically completed.

The whole culture is added to the grain mash, previously cooked, and cooled to conversion temperature at the rate of approximately 3.0 gallons per 56-pound bushel of corn. After primary conversion, the grain mash is cooled and inoculated with yeast. The alcoholic fermentation is finished in about 72 hours. Yield of alcohol from grain converted with mold amylases is not less than that obtained when barley malt is employed as the saccharifying agent.

The procedures for developing laboratory cultures of Aspergillus niger are a little more precise. A *slant* mentioned in the following directions is merely a culture medium solidified obliquely in a tube so as to increase the surface area. You will notice that the stuff grows in quantum leaps of a hundred. It's a wonder Hollywood hasn't made a movie about *The Fungus That Ate California* or something. It could be another film classic like *Attack of The Killer Tomatoes*.

Don't panic over what a *carboy* is; I'll get to it directly. The following is taken directly from a U.S. Department of Agriculture Technical bulletin.

PREPARATION OF MOLD INOCULA

The methods and techniques employed for developing laboratory cultures of Aspergillus niger, NRRL 337, were those recommended and established by the Northern Regional Research Laboratory. The general procedure used to increase the quantity of culture and details of the various laboratory processes follow:

parent or stock culture (slant)→ slant → flask → flask
carboy (2.5 gallons)→ seed tank (250 gallons)
amylase plant fermentor (about 25,000 gallons)

Sporulated cultures of Aspergillus niger, NRRL 337, obtained from the Northern Regional Research Laboratory, are stored in a refrigerator. To initiate the production of an inoculum, a few spores are transferred from the parent culture to a test tube slant. The culture medium and method of preparation are as follows:

Distillers' thin stillage fortified with solubles or sirup to a solids content of 5 percent.

2 percent corn starch.

2 percent agar.
Sodium hydroxide to adjust pH to 6.0.
Plug tubes with cotton and sterilize 30 minutes at 250°F (15 pounds to the square-inch gauge).

Be a little careful with the sodium hydroxide, or NaOH. You can only adjust pH up, not down, with it. What NaOH does is fall apart into Na and OH molecules. The OH soaks up the H's in your solution and makes H_2O's out of them. The acidity of a solution is determined solely by how many H's—or hydrogen atoms—are floating around loose.

Back to the directions.

This transfer and all others must be conducted aseptically because no degree of contamination can be tolerated in laboratory or seed cultures. After the slant has been incubated for 24 hours at 87° F, a part of the vegetative growth produced is transferred to a 1-liter flask containing 200 milliliters of medium of the following composition:

Distillers' thin stillage fortified with solubles or sirup to a solids content of 5 percent.
2 percent cornstarch solubilized with 0.1 percent barley malt.
0.5 percent calcium carbonate.
Sodium hydroxide, if necessary, to adjust pH to 5.0 to 6.0.
Plug flasks with cotton and sterilize 30 minutes at 250° F (15 p.s.i.g.).

In order to supply oxygen to the organism, the flask is shaken in a room or cabinet in which the temperature is thermostatically controlled at 87° F. The speed of the shaker is approximately 45 cycles per minute. The flask is incubated and agitated for 24 hours. Then a small part of its contents (5 milliliters) is transferred to a second 1-liter flask which contains 200 milliliters of medium of the same composition. This flask is shaken at 87° F for 24 hours. At that time, a vigorous and healthy growth is obtained.

The entire contents of the second flask are employed to inoculate 10 liters of sterile medium which is contained in a 4.5-gallon carboy. The medium is of the same composition as that used in the flasks; however, because of the large quantity of liquor, the bottle and contents are sterilized for 60 minutes at 267° F (25 p.s.i.g.). The inoculated medium is aerated vigorously by the passage of air into the carboy through a glass or copper tube. Air for this purpose is sterilized by filtration through tubes packed with cotton or glass wool. In addition to the air inlet, the carboy is equipped with an air vent and a tube through which the contents of the bottle are transferred to the dona tank. After aeration and incubation at 85° to 90° F for 24 hours, the carboy culture may be used to inoculate the dona, or seed tank in the fungal amylase plant. Details of the methods and equipment employed to transfer culture liquors from flask to carboy and carboy to seed tank, and for the aeration of the carboy culture, are as follows.

PRODUCTION AND TRANSFER OF INOCULUM

The first steps in the preparation of inoculum involve culturing of the mold on slants. A few spores are transferred from the parent or stock slant cultures to a second slant which is incubated to propagate the fungus. And a small portion of the vegetative growth is then transferred to 200 ml. of broth in a 1-liter Erlenmeyer flask. No special techniques or equipment are required for these operations; standard bacteriological procedures are quite satisfactory.

After the flask-culture is incubated on a shaker for 24 hours, part of the culture liquor is used to inoculate a second flask identical in size to the first flask, and which contains the same quantity of medium. A pipette is used to make this transfer and regular aseptic precautions are observed. The second flask, like the first, is simply plugged with cotton (A of Fig. 8-1).

After incubation, the total contents of the second flask are used to inoculate a carboy of sterile medium. The carboy with special fittings is represented by A of Fig. 8-2. It is equipped with a tight-fitting stopper which has openings for three tubes. Tube (a) is a vent and serves also as the inoculum inlet; tube (b) is the line through which the carboy culture is discharged; and air is passed into the culture through tube (c). The length of tubing (e) is packed with cotton or glass-wool. Air that is passed into the culture during fermentation is sterilized by means of this filter.

To prepare the carboy culture, the bottle is charged with medium and closed with the stopper that is complete with appurtenances. The stopper is fastened in place securely by means of a wire clamp. Vent-tube (a) is plugged with cotton and the special fitting (d) is plugged and wrapped with cotton. Pinch-clamps (f) and (g) are closed. The entire unit is then sterilized in an autoclave for 60 minutes at 267° F (25 p.s.i.g.).

Inoculum is transferred from the second shaker flask into the carboy. For this purpose, the apparatus as illustrated by B of Fig. 8-1 is used. The glass tube (p) is protected by means of the test

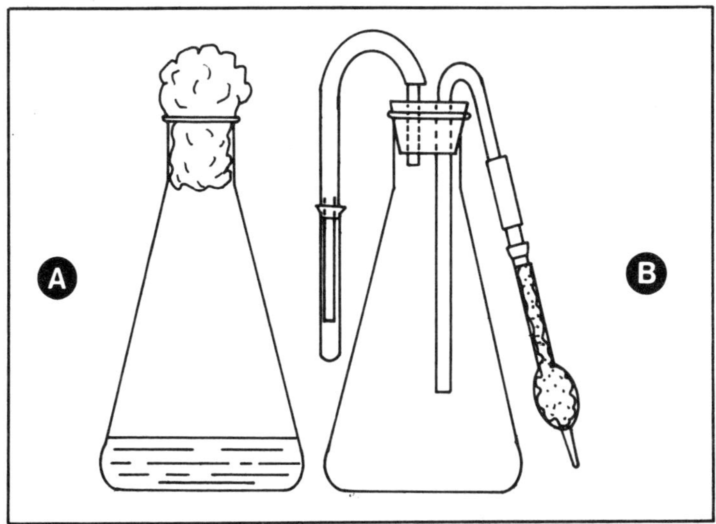

Fig. 8-1. A Shaker-flask (A) and an apparatus for transferring culture (B).

tube (m) which fits snugly onto the rubber tubing. The filter tube (n) is packed with cotton. The assembly, including the stopper, is placed in an empty 1-liter flask and the stopper and top of the flask

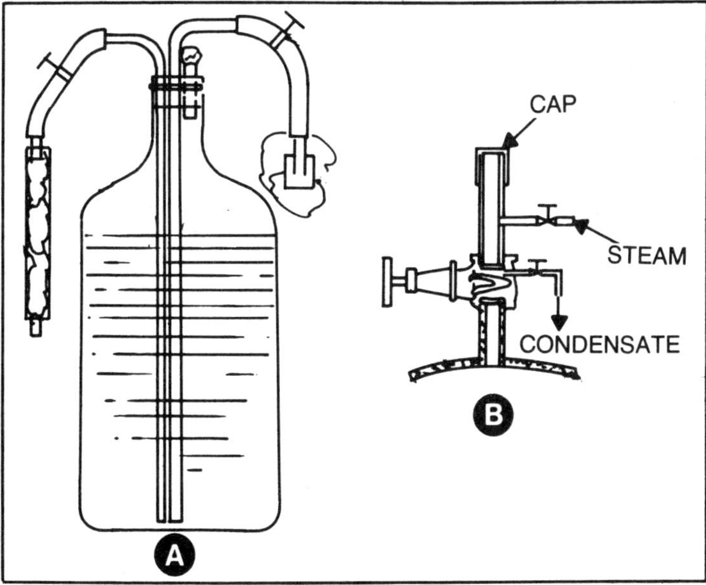

Fig. 8-2. Details of a Carboy (A) in which a culture is grown and an inoculum valve on a seed tank (B).

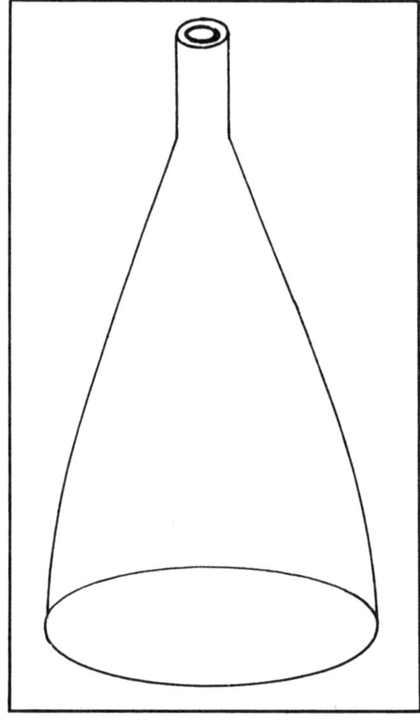

Fig. 8-3. An Erlenmeyer flask.

are wrapped in cotton. The whole unit is then sterilized in an autoclave.

To transfer inoculum from a shaker flask to a carboy, the assembly is shifted from the empty flask to the shaker flask; the test-tube is removed from the glass delivery tube; the cotton plug is removed from the vent opening in the carboy; and the delivery tube of the flask assembly is inserted into the carboy vent. These operations are conducted aseptically.

The flask is then inverted and the culture flows into the carboy. If clumps of mycelium plug the delivery line, air is pumped into the flask by means of an aspirator bulb attached to the end of the filter—in order to force the culture through the tube. After the transfer is made, the carboy vent is again plugged, an air line is attached to the end of the air filter (e), pinchclamp (f) is opened, and the flow of air into the carboy is begun.

Inoculum is transferred from the carboy into a seed tank through fitting (d), (A of Fig. 8-2) into the inoculum valve on the seed tank. B of Fig. 8-2 is a sketch of the valve assembly. Except when the tank is being seeded, the inoculum valve is closed, the

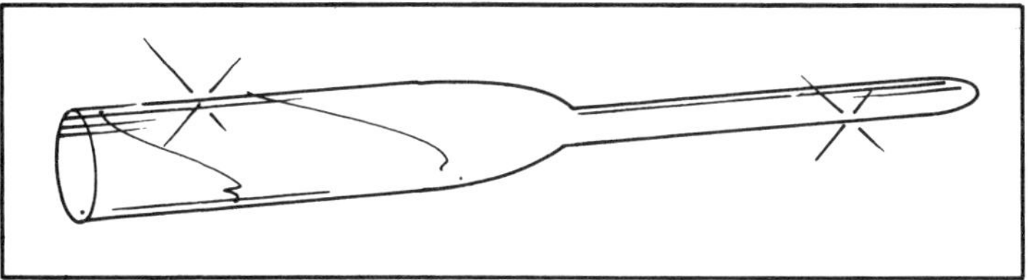

Fig. 8-4. A pipette.

drip valve is open, and sufficient steam is passed into the pipe nipple above the valve so that vapors escape from around the cap and through the drip line. At the time of inoculation, the cap is removed, the cotton covering and plug removed from fitting (d) of the carboy, and the adaptor inserted into the opening of the inoculum valve assembly. The inoculum valve is opened, the steam and drip valves are closed, and pinch clamp (g) is opened. Air is passed into the carboy through the filter (e), the vent (a) is closed by covering it with a finger, and the pressure which develops forces the culture liquor into the seed tank.

Figure 8-3 illustrates an *Erlenmeyer flask*. It is simply a jar with a wide bottom and narrow mouth that is commonly used in chemistry labs. They are more difficult to tip over than a normal jar.

Figure 8-4 illustrates a *pipette*. If you want to pick something up in tiny amounts you must dip the small end into your project, put your thumb over the large end, and pull it out. Take your thumb off to drop it into the next area. These are also known as "Pasteur pipettes." Both the flasks and the pipettes are made of glass and are available from any chemical supply house.

An *autoclave* is merely an extremely strong steam boiler that is used for sterilization and other projects. The object to be sterilized is, in effect, put *inside* the boiler with just a little water and then heat is applied. The water turns to steam which, in turn, creates pressure in addition to the heat already supplied. One of the unusual things you can do with an autoclave and a little coal or hydrogen gas is turn animal waste into diesel fuel.

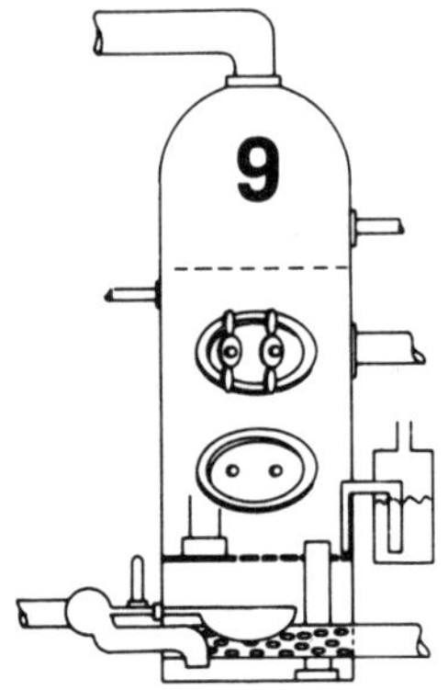

Starch Fermentation

Until World War II, practically all alcohol used industrially in this country was produced by the barley malt conversion of starch to sugar and fermentation of sugar by yeast. Probably more progress would have been made except for the halting of research in the area due to the passing and enforcement of prohibition, 1920 to 1933. The alcohol business was given added inpetus with the beginning of World War II because one of the byproducts of butanol production—acetone—was desperately needed for the manufacture of smokeless gunpowder.

ALCOHOL PRODUCTION

During and after World War II, the production of synthetic alcohol from petroleum overtook and surpassed the fermentation alcohol industry. At the time, the bulk of the industrial alcohol in the United States was produced from molasses. Drinkable alcohol from molasses is called rum.

Another reason alcohol production took top priority during World War II was because industrial alcohol was needed for our synthetic rubber program. Barley malt was just too scarce and expensive for making rubber tires. There had to be a cheaper, more plentiful substitute.

The production of alcohol from starch is a simple three-step process:

—Convert starch to sugar.
—Convert sugar to alcohol.
—Distill the alcohol.

The first step caused all the problems. It isn't necessary for molasses because that substance is already a sugar. It is a must for all starch products. Step number one is where barley malt or a substitute must be used.

In 1946, some technicians in Peoria, Illinois prepared a fungal amylase agent that could be used for converting starch to sugar. Fungal simply means consisting of fungi—parasitic plants that lack chlorophyll. An amylase is any of the enzymes that accelerate the hydrolysis—*hydro* for water, *lysis* for splitting—of starch. A starch is merely a group of sugar molecules glued together. The yeasts are simply not strong enough to attack a starch molecule and convert it to alcohol and carbon dioxide.

The fungal amylase was made by fermentation of stillage-corn medium with a mold—called Aspergillus niger, NRRL 337. A lot of the companies in the business selling enzymes with "trade names" are merely using this same fungal amylase. If you're going for volume production, it is a lot cheaper to manufacture (or should I say breed?) your own enzymes than it is to lay out money for someone else's trade name. It is rather easy to cost yourself 10 cents a gallon more if you don't grow your own. On a small scale, 10 cents a gallon doesn't make much difference, but on a large operation it could cut your profit down to almost zero. In a 14-gallon per minute 5-foot diameter column, you simply couldn't afford it.

If you buy your enzymes from someone advertising in *Gasohol U.S.A.* (Box 9547, Kansas City, Missouri 64133) or a similar publication, the following method is all you have to do to get your starch ready to distill. Grind the starch (such as corn) up in a hammer mill. This exposes as much of the starch as possible to enzyme action. The next step is to simply dump the ground starch into boiling water. In the case of corn, there should be 28 gallons of water for every bushel. With no water—or too little—what you wind up with is mechanical entrapment and no way for the enzymes to go to work on the starch.

In case you're wondering what an enzyme is, it comes from two Greek words: *en* simply meaning "in" and *zyme* meaning yeast. Enzymes are biochemical catalysts that are soluble in water. A catalyst is merely a substance that speeds up a reaction, but doesn't enter into the reaction itself. An example is the catalytic converter on your car. Supposedly, a gram of plutonium imbedded in an aluminum oxide base cleans up some of the exhaust gas coming out of your car without being used up itself. I say supposedly because catalytic converters only work on paper and in government documents. To my knowledge, no one has ever actually seen one work.

There are two types of enzymes. *Endo* or *intracellular enzymes* work within a biological cell, involve oxidation-reduction reactions, and involve energy changes. The type of enzyme we are concerned with here is an *exo* or *extra-cellular enzyme.* These appear in the medium surrounding the organisms. Because such reactions involve the addition of water, the enzymes are also called *hydrolases.*

Back to the ground-up corn. Boil the corn for 15 to 30 minutes. You can then either let it cool naturally or pump cooling water in. The cooling water does not have to be mixed with your mash or ground-up corn and water. A simple copper coil inside your cooker with cold water running through it will easily carry away enough heat.

When the temperature reaches 150° to 170°F it is time to dump in your enzymes. The ideal temperature is 152°F. It is also about the top level temperature for barley malt. Fungal amylase appears to be more heat resistant.

At these temperatures, *saccharification*—the changing of starch to sugar—takes place. Once this mash has cooled to 120°F, the conversion can be considered complete. To see if it is complete, there are two ways you can test it. One method is the iodine test.

Take a small sample of mash and put it in a cup. Using an eyedropper, administer a few drops of iodine. If no color results you have a complete conversion. If the liquid turns blue there is still starch present. The problem here is that only "all or nothing" is registered. A 50% conversion will turn just as blue as a 90% conversion. If you are running a backyard operation and your yield of alcohol per bushel is low, this might be one of your problems.

BALLING HYDROMETER

You can get a more accurate measurement with a *balling hydrometer* (Fig. 9-1). A regular hydrometer measures specific gravity of a fluid and is used to determine things like the amount of water in alcohol or the amount of sulfuric acid in a battery. A balling hydrometer measures sugar content in liquid.

A balling hydrometer is shaped something like a vertical torpedo with a long, skinny tail. The lower tip is weighted. On the stem, the per cent sugar (Brix or Balling) reads from 0 to 24 with the numbers decreasing toward the top. The main body contains a thermometer with a scale reading from 30°F to 140°F. A series of red numbers from red 1.0 reading down to .6 are printed on the right side of the thermometer. That is a correction scale. The tip of the balling hydrometer, or *sacchrometer* (sugar hydrometer), is weighted to keep it upright in the liquid.

The way you read a balling hydrometer is as follows. An "O" on the stem means there has been a complete conversion of starch to sugar. At least all that is available has been converted. In real life this seldom happens. The "O" is predicated on an ideal temperature of 60°F. That is why the thermometer squats in the bottom section of the balling hydrometer.

If you pull the hydrometer out of your mash and the temperature reads "80°F, then you look parallel to the "80" and find ".8" (the conversion factor) in red to the right. Now look to the stem (the number that is visible at the liquid level in your mash). Let's say that the number is "8." Add .8 to 8 for a corrected reading of 8.8 at 80 degrees. At 70 degrees, the corrected reading would be 8.4. At 60 degrees it would just be 8. At 40 degrees, it would be 7.4. Below 60°F you subtract.

If you measure something as thick as molasses, the hydrometer will simply flop over on its side because the syrup is too thick. The way you get around that is to dilute the liquid and then multiply the balling scale by the dilution factor.

In plain English, if you have a quart of molasses and dilute it with 3 quarts of water you have a dilution factor of four. Multiply the four by whatever is even with the liquid level on the balling scale. If it reads "8" on the balling scale, then 8 x 4 is your true reading: or 32.

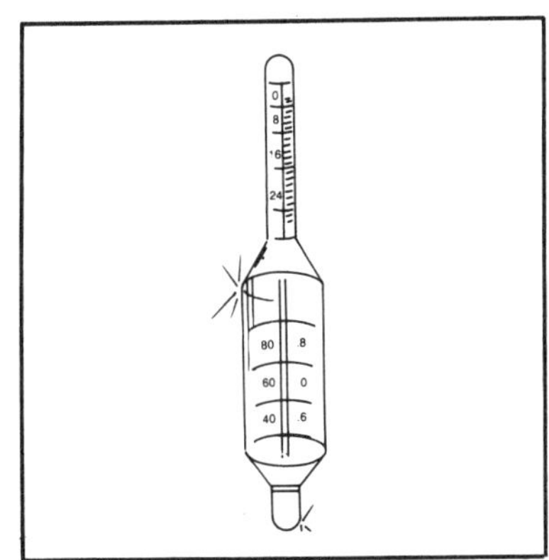

Fig. 9-1. A balling hydrometer.

WORKING THE MASH

Once you are satisfied with your sugar conversion, it is time to work the mash further. The mash must again be cooled. This time it must be cooled to below 90°F. At this point yeast is added. Normally, it is just plain baker's yeast like the type you buy in the supermarket. You might want to add coolers at this point to keep the temperature at 86°F (an ideal temperature for fermentation). Don't stir it, just let it alone for two or three days. Carbon dioxide bubbles should form and rise to the surface. The mash is now ready to distill.

A word of caution may be in order here. On one of my trips to Nebraska, I inspected a fermentation operation in the basement of a man's house. I say "fermentation operation" because he was so convinced his mash wasn't working off right that he hadn't even built a still. He wanted to first get the fermentation procedure down pat.

His wife mentioned that there had been something wrong in the house since he started experimenting. Everyone living in the house was exhausted. When they got out of bed in the morning, they could hardly drag themselves around. Only after a period of time away from the house did their energy return.

We entered his basement and I immediately saw the problem. Carbon dioxide bubbles the size of baseballs percolated up from his mash. He handed me his hydrometer to measure the mash. His hydrometer had a hole in it. It leaked. And it was a whiskey hydrometer. It simply wouldn't work in anything except high-proof alcohol.

What he was doing to himself and his family was producing so much carbon dioxide that it was flooding his house and displacing a good deal of the normal atmospheric oxygen. His whole family was starved for oxygen.

Carbon dioxide is not harmless. It is nowhere near as deadly as carbon monoxide, which causes problems in the bloodstream, but it can still kill you. Carbon dioxide is heavier than air, settles to the ground, is odorless, colorless, and tasteless, and it can occur in sufficient quantities to suffocate you unaware.

If you must experiment in your basement or some other enclosed place, cover your fermenting barrels (or tanks) and run vent lines well outside your house. The wind will disperse the carbon dioxide and render it harmless. The fermenting process does not require oxygen. You don't have to worry about obtaining fresh air for your process.

Commercial Enzyme Preparation And Troubleshooting

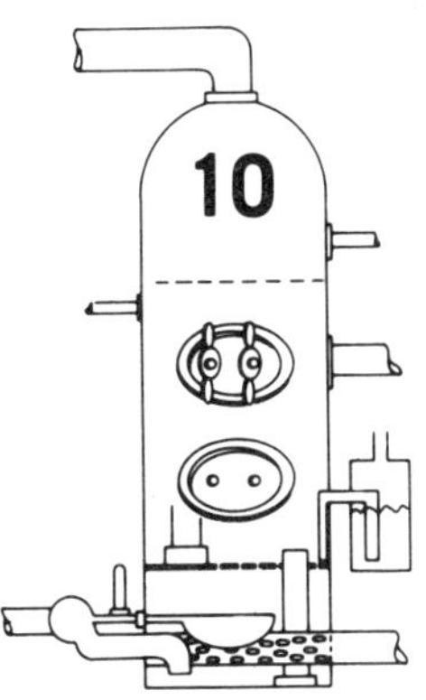

The following is a reproduction of a technical bulletin put out by the Department of Agriculture. Because it is a technical bulletin and because most of the folks interested in alcohol do not speak the technical jargon, I have taken the liberty of interspersing the original bulletin with my own explanations.

The company mentioned in this bulletin is not interested in helping neophytes. It's unlikely anyone working in the office would be able to help you anyway. A friend of mine called them and was informed that Grain Processors was getting 200-proof alcohol out the top of a single column without the use of a drying agent. Typical office clerk, I guess.

This is a bit technical and must be read through numerous times before it all sinks in to stay. Where possible, I have added explanations to simplify things.

DESIGN, INSTALLATION, AND OPERATION OF FUNGAL AMYLASE UNIT AT PLANT OF GRAIN PROCESSING CORPORATION, MUSCATINE, IOWA

In the pilot-plant work on the fungal amylase process conducted at the Northern Regional Research Laboratory, mold fermentations were carried out in copper 800-gallon tanks, and alcoholic fermentations were made in steel fermentors, each of which had a capacity of 4,000 gallons. During the experiments, approximately 300 gallons of fungal amylase was prepared per batch and a portion of this was employed to convert about 1,400 gallons of grain mash (40 bushels of corn). It was ascertained that 7 to 10 percent of the mold agent, based on the final volume of the grain mash, was sufficient for conversion, and that when batch cooking and conversion were used the fungal amylase could simply be substituted for a slurry of malt with no other change in mashing operations.

The effect of rate of aeration during fermentation on the enzyme potency of the fungal amylase liquor had been investigated. Although one-eighth volume of air per volume of medium per minute appeared to be sufficient for the preparation of a satisfactory product, it was recommended that an aeration rate of one-quarter be employed for the plant operation, as the minimum rate appeared to be a function of the shape of the vessel and the degree and type of agitation provided. A satisfactory medium had been developed and tested during the pilot-plant and laboratory investigations. This medium, to be used in the plant, consisted of thin stillage with a solids content of about 5 percent, supplemented with 1 percent of ground corn and 0.25 to 0.50 percent of calcium carbonate.

It had been determined that the yield of alcohol from corn converted with fungal amylase was no less than, and in some instances significantly higher than, the yield with malt as the saccharifying agent. Much information had been obtained on the pure culture techniques needed in the large-scale production of fungal amylase. All available results and engineering data of the Northern Regional Research Laboratory were placed at the disposal of the operators of the Grain Processing Corporation to be used as bases for the design and operation of a large installation.

It had been decided that sufficient fungal amylase was to be produced per run to convert enough grain to set two large fermentors in the distillery. As the capacity of each alcohol fermentor is 100,000 gallons, and approximately 10 percent fungal amylase was to be employed for conversion, the required capacity of the fungal amylase plant was 20,000 gallons per run.

The design of the enzyme plant was influenced to a major extent by the type of tank or fermentor that could be obtained readily. The closed tank which was purchased is constructed of lightweight steel plates, and is cylindrical, with flat top and conical bottom. The tank has a total capacity of 30,000 gallons, and cannot be pressure-sterilized because of its construction. Although a simple installation was desirable for the experiments, it obviously was impossible to sterilize the medium batchwise in the fermentor; hence, a continuous cooker was provided and the tank was cleaned by mechanical and chemical means before each run.

Figure 10-1 is a diagrammatic flowchart of the process as installed and used. Each item of equipment shown on the flowchart is numbered and corresponding numbers are used in the following description.

(1) Mix Tank. This unit is an open cylindrical tank equipped with a side-entering agitator. Its total capacity is 500 gallons. It is used for blending supplementary materials with the stillage prior to sterilization of the medium.

(2) Cooker Feed Pump. The pump has a rated capacity of 50 g.p.m. against a head of 75 pounds to the square-inch gauge and is driven with a 15-horsepower motor. It is employed to pump stillage with added ingredients through the cooker and cooling system and into the fermentor.

(3) Cooker. This unit consists of a 1½-inch jet heater and 40 feet of 6-inch standard iron pipe. The medium is heated instantaneously in the jet heater and is kept at the elevated temperature during its passage through the section of pipe.

(4) Cooler. Hot, sterile medium is cooled continuously to fermentation temperature in this equipment. It consists of two double-pipe or concentric-pipe heat exchangers connected in series. The total surface for cooling in both units is 235 square feet.

(5) Fermentor. This cylindrical closed tank is constructed of welded steel plate, with a flat top and conical bottom. It is 20 feet in diameter and 12 feet high on the straight side, and has a total capacity of approximately 30,000 gallons. The fermentor is equipped with a top-entering mixer, which is driven at the rate of 32 revolutions per minute by means of a 75-horsepower motor with chain and gear reduction. The mixer consists of one turbine-type unit, 84 inches in diameter; the tank is equipped with four vertical baffles to reduce swirling of the medium during agitation. Air is introduced into the fermentor by four 2-inch pipe lines, the discharge ends of which are just below the turbine agitator. All piped openings into the tank are protected against the accidental passage of contaminants by double valves, and by steaming the section of pipe between the valves. Air is exhausted from the tank through a 6-inch pipe welded in the top of the tank. This line is piped in the shape of an inverted "U" to prevent contaminating moisture or dust from entering the vessel.

(6) Seed Tank. This is a closed cylindrical steel tank, 3½ feet in diameter and 5 feet high, with flat heads; its total capacity is approximately 360 gallons. It is equipped with at top-entering mixer with 2-propeller agitators. Air is sparged into the fermentor

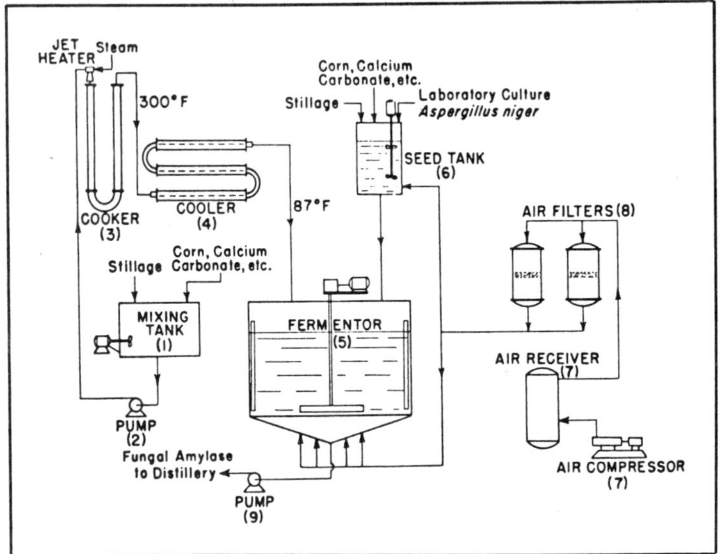

Fig. 10-1. A flowchart for the production of fungal amylase.

through a ¾-inch tee near the bottom of the vessel under the lower propeller. All piped openings into the tank are steam-sealed. Air is exhausted from the unit through a section of pipe, which is trapped by an inverted "U" bend to prevent the entrance of contaminated moisture or dust into the tank. A flow-meter is connected to the discharge end of the vent line and used to measure the volume of air that is passed through the medium.

(7) Air Compressor. This is a single-stage, reciprocating compressor with a rated capacity of 1,000 cubic feet per minute at 20 pounds to the square-inch gauge. It is driven by a 75-hp. motor. An air receiver, 425 cubic feet in capacity, is provided. It was determined that the actual capacity of the compressor was about 650 c.f.m. at 20 p.s.i.g. and 900 c.f.m. at 15 p.s.i.g.

(8) Air Filters. Two filters are provided. Each consists of a tank 3 feet in diameter by 4 feet high (shell of Bowser filter), with a removable top head and equipped with a single horizontal support for the filter medium. This is nonabsorbent cotton, of which 16 1-inch pads compressed to a thickness of approximately 4½ inches are used. The tanks are constructed to withstand internal pressure; hence, they can be sterilized with steam at 15 pounds to the square-inch gauge.

(9) Fungal Amylase Pump. This is a centrifugal pump with a capacity of 25 gallons per minute against a head of 50 feet. It is driven by means of a 5-horsepower motor, and is used for transferring fungal amylase from the fermentor to the distillery.

The final assembly and installation of the equipment were made with knowledge that fungal amylase must be produced by pure culture fermentation. Items of equipment which must be sterile or free of contaminants during all or some portion of the fermentation cycle are the cooker, cooler, fermentor, air filters, seed tank, and pipe lines that connect these units. Therefore, provisions were made to sterilize the equipment and lines with steam at a pressure of 15 pounds to the square-inch gauge when possible. Obviously, in an installation of this type, the equipment and pipe lines must be arranged so that they can be cleaned and kept clean if steam sterilizer is to be completely effective. For example, if medium can enter the air line and particles of corn are deposited in the pipe between the air filter and the fermentor, these pockets of organic material may become centers of infection and cause contamination of the medium in later runs.

Although it appears improbable, it is possible that infectious material might leak past the gate of a closed valve in a pipe line connected to the fermentor or seed tank. For this reason sample outlets, mash inlet lines, drain outlets, etc. are double-valved and a steam seal is provided between the valves. In general, this method should be used as a safety measure in all piped connections between the fermentor or seed tanks and the appurtenances which are or may be non-sterile.

The fungal amylase plant was erected and operated as an outdoor unit. Subsequently, however, it was enclosed in a frame building so that operations could be continued during winter.

Plant Operation. Thin stillage was pumped from a storage tank in the feed recovery house of the distillery to the mix tank in the fungal amylase plant. The solids content of the stillage was in most cases about 4 percent and it was increased to 6 to 8 percent by fortification of the stillage with distillers' sirup (evaporated stillage) that contained approximately 25 percent solids. Ground corn, calcium carbonate, and a solution of sodium hydroxide were added proportionally to the fortified stillage in the mix tank. These ingredients were added in quantities sufficient to give the medium a composition of approximately 1 percent corn and 0.25 to 0.50 percent calcium carbonate. Its initial pH was 5.0 to 5.5. A small amount of ammonium bifluoride also was added to the stillage at this time, for the purpose of inhibiting bacterial growth in the medium if it became contaminated during fermentation.

In a typical plant run (Experiment 12B), about 25,000 gallons of medium was prepared. To produce the liquor, fortified thin stillage was supplemented with 2,000 pounds of ground corn, 1,000 pounds of calcium carbonate, 325 pounds of sodium hydroxide, and 37 pounds of ammonium bifluoride. The pH of the medium after sterilization was 5.0. Based on the volume at this time, it contained 0.96 percent of corn 0.48 percent of calcium carbonate, and 0.018 percent of ammonium bifluoride.

Stillage, mixed with corn and the other ingredients, was withdrawn continuously from the mix tank and pumped through the cooker and cooler into the fermentor. In preparation for this operation, the cooker and cooler had been sterilized with steam under pressure, and the fermentor had been cleaned chemically with detergents and antiseptics and then steamed at atmospheric pressure. The temperature of the medium was increased instantaneously to 290° F by mixing it with steam at 55 pounds per square-inch gage in the jet heater, and the medium was retained at this temperature for 6 minutes during its passage through the cooker. The rate of pumping was 30 to 35 gallons per minute. The hot medium passed from the cooker through an expansion valve to the cooler, where its temperature was reduced to 87°F, and then to the fermentor. A total time of about 12 hours was required to sterilize 25,000 gallons of medium.

From 24 to 30 hours before the large fermentor was charged with sterile medium, the seed tank had been inoculated. The seed medium was prepared and sterilized batchwise. Approximately 250 gallons of fortified thin stillage was pumped into the seed tank, the agitator started, and ground corn, calcium carbonate, sodium hydroxide, and ammonium bifluoride were added. The volume of medium at this time was about 260 gallons. In experiment 12B, the quantities of ingredients used were as follows: 20 pounds of corn, 7.5 pounds of calcium carbonate, 2.0 pounds of sodium hydroxide, and 140 grams of ammonium bifluoride. The composition of the medium before sterilization was 0.92 percent corn, 0.35 percent calcium carbonate, and 0.014 percent ammonium bifluoride, and the pH after sterilization was 5.6.

The medium was sterilized in the seed tank by double cooking, with an intervening rest period. After the medium was cooked with open steam for 60 minutes at 240°F, it was cooled to 90° to 100°F and kept at this temperature for several hours. The medium was recooked for 60 minutes at 270°F, then cooled to 87°F and inoculated with 10 liters of a laboratory culture of *Aspergillus niger*, NRRl 337. Hot medium was cooled by means of a spray of cold water on the outside of the tank. Cold water or warm condensate was sprayed on the tank to control the temperature of the medium at 84° to 88°F during fermentation.

Sterile air was supplied the seed fermentor while the cooked medium was being cooled. The contents of the tank were under atmospheric pressure only when the inoculum was transferred into it. During fermentation, the rate of aeration with sterile air was about one volume per volume of medium per minute.

After the seed culture had been incubated for 24 to 30 hours, it was transferred to the large fermentor. This was done by closing the vent valve of the seed tank, while aeration was continued, to increase the pressure in the vessel to 12 to 15 p.s.i.g. which was sufficient to force the culture into the large fermentor through a connecting pipe line. The transfer line had been cleaned and was steamed until used. The large fermentor was inoculated while it was being charged, and usually when it contained about 8,000 gallons, which was enough medium to cover the turbine agitator.

Aeration in the large fermentor was begun immediately after it had been steamed and just before sterile medium was pumped into the tank. In this way, the pressure inside the vessel always was slightly greater than atmospheric, a condition which is recommended for pure culture fermentation. Aeration of the medium in the large fermentor was conducted at the rate of 650 to 950 c.f.m., which corresponded to a volumetric rate of 0.20 to 0.28 volumes of air per volume of medium per minute, when 25,000 gallons of liquor was fermented. The temperature of fermenting medium in the large tank was adjusted by means of a spray of cold water or warm condensate on the outside of the vessel. This temperature ranged usually between 85° to 89° F.

After fermentation for 48 to 60 hours, the fungal amylase liquor was ready for use. At this time, it was pumped to the distillery as needed during mashing operations there.

USE OF FUNGAL AMYLASE

In order that the reader might fully comprehend the use of fungal amylase in the distillery, the system of mashing employed at the Grain Processing Corporation is explained here in detail. Figure 10-2 is a diagrammatic flowchart of the process used at Muscatine to prepare grain for fermentation with yeast. Corn is ground in hammer mills and the ground grain is mixed with water and stillage in the premix tank or precooker. The slurry is pumped

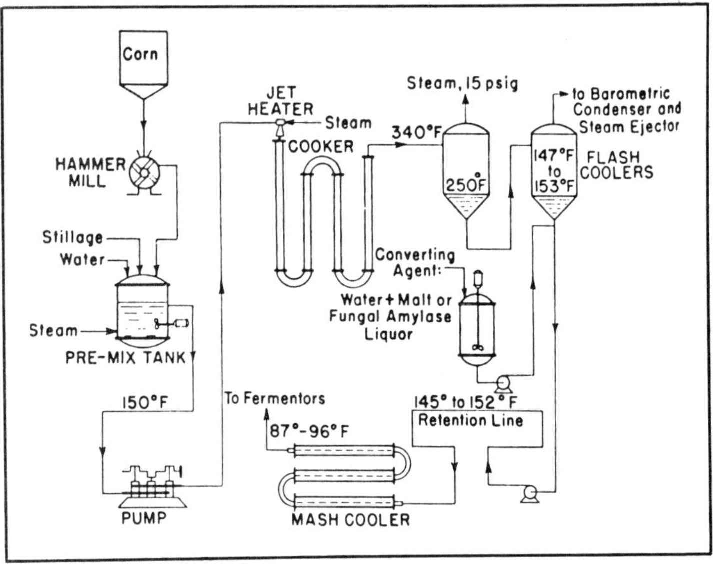

Fig. 10-2. A flowchart showing how grain is mashed in the alcohol plant of the Grain Processing Corporation.

from this vessel to a continuous cooker. A jet heater at the entrance to the pipe cooker instantaneously heats the mash to 340° F and it remains at this temperature for about 5 minutes during its passage through the retention leg of the unit. The hot mash is cooled to 250° F in a pressure flash cooler; then to 147° to 154° F in a vacuum flash cooler. The converting agent is introduced into the mash at this point, and the mixture is pumped through a length of pipe in which the mash is retained for 5 to 6 minutes at the conversion temperature that is usually 145° F when malt is used. This is referred to as the period of primary conversion. Mash passes from the conversion line to the mash coolers, where its temperature is reduced to about 90° F and then to a fermentor.

When malt is used as the converting agent, it is ground, mixed with water in a malt slurry tank, and the mixture is pumped into the mash stream. To change from malt to fungal amylase was a simple operation; the mold liquor was pumped from the fungal amylase plant to the malt slurry tank, from which it was introduced in the regular way into the mash stream.

Several purposes are accomplished when grain is mashed. In grinding grain, the starchy portions of the kernel are exposed, but only a small part of the starch is available to enzyme action. When an aqueous slurry of ground grain is cooked, the starch granules swell and rupture; the starch becomes gelatinized and thus open to attack by enzymes. When the cooked mash is cooled, it becomes very viscous, and if the concentration of grain in the mash is normal (20 to 22 percent) it is impossible practically to transport the material mechanically; that is, by pumps and through pipe lines. Also, at this time, only a small amount of fermentable sugar is present in the mash, and some sugar is required for immediate use by the yeast in the subsequent alcoholic fermentation. To liquefy or thin the mash, and to degrade a portion of the starch to dextrins and fermentable sugars, *amylolytic enzymes* are added to the cooked and partially cooled material. When mashing is continuous, the conversion temperature employed must be high enough to saccharify the starch and to thin the mash to the desired extent in a few minutes. The temperature must not be so high as to cause destruction of the enzymes.

This initial treatment of starch at an elevated temperature with enzymes is called primary conversion. The mash has been only partially converted. Saccharification of the residual starch and of the dextrins takes place slowly at the temperature of and during the alcoholic fermentation. This is the period of secondary conversion.

In the pilot-plant work at the Northern Regional Research Laboratory, corn mashes were cooked batchwise in all instances. No difficulties were encountered; both liquefaction of the mash and saccharification of the starch were satisfactory. However, because the conversion step was a batch operation, time was not a factor. Experimental results indicated that fungal amylase did not thin a grain mash as rapidly as malt under conditions usually employed when malt was used as the converting agent. It was beneficial, in the tests at Muscatine, to increase the conversion temperature with fungal amylase to 152° F in order to enhance the liquefaction power of the material.

The alcoholic fermentation of mash converted with fungal amylase was conducted in the regular manner. Each fermentor in the plant of the Grain Processing Corporation has a capacity of 100,000 gallons, and in these experiments one tank was charged with 2,700 to 3,500 bushels of corn, the amount charged depending on the concentration of grain in the mash. One batch of fungal amylase liquor was used to convert mash for two plant fermentors. Temperature of the fermenting mash was adjusted when necessary by circulation of the liquor through a heat exchanger at the base of a fermentor. The yeast inoculum was prepared from malt-corn mash which was lactic-soured. Most alcoholic fermentations of mash converted with fungal amylase were in process over a week end; hence, a beer was 4 or 5 days old at the time it was distilled.

At the time these contractual experiments were conducted at Muscatine, the plant was engaged in the production of neutral spirits for use in beverages. The distillation unit, a four-column still, gave a product of satisfactory quality as judged by organoleptic tests. The operation of the columns was not altered for the distillation of fungal amylase beer. Because neither the beer well nor the distillation unit was empty when it was time to distill mold beer, it was impossible to make a sharp separation between malt spirits and mold spirits. Samples taken during the latter part of a distillation were considered bo be representative of mold alcohol.

Byproduct feeds were recovered at the plant by the regular methods. Whole stillage was screened and the oversize particles of grains pressed and dried to make distillers' light grains. The thin stillage that passed through the screen was concentrated by evaporation to a sirup containing about 25 percent solids. Sirup was pumped to a double drum dryer where distillers' dried solubles were produced. During the experiments, some distillers' dark grains were made. In this case, the dried light grains were mixed with sirup and the mixture dried in a rotary unit. Again, there was some difficulty in sharply separating mold and malt byproduct feeds because of the constant flow of material through the plant.

PRODUCTION AND UTILIZATION OF FUNGAL AMYLASE

A total of 31 experiments on the production of fungal amylase was conducted during the course of the work at Muscatine. Of these runs, the first 15 were preliminary and served mainly to familiarize the operators with the process and with the techniques involved. The fungal liquors produced in the preliminary runs were unsatisfactory in most instances; only one batch was used in the distillery to convert mash.

Information on the last 16 experiments is given in Table 10-1. Approximately 25,000 gallons of fungal amylase liquor was prepared in each of these runs, and with the exception of material made in experiments 18B and 23B, each batch of mold liquor was used in the distillery for the conversion of 5,000 to 7,000 bushels of grain.

Fortified thin stillage was employed as the principal substrate of the medium in all mold fermentations. Because in regular operations grain was converted with malt, stillage most easily

Table 10-1. Data on Production of Fungal Amylase Liquor on a Semiplant Scale.

Run number	Composition of medium [1]				Time of fermentation	Analysis of fungal amylase liquor				
	Type of stillage	Corn	Calcium carbonate	Acidity		Saccharification value, conversion	Alpha-amylase	Maltase	Acidity	
									Titratable [2]	Final
		Percent	*Percent*	*pH*	*Hours*	*Percent*	*Units (30° C.) /ml.*	*Units/ml.*		*pH*
9B	Corn-malt	0.96	0.48	5.5	52	24.0	12.0	4.0	5.2	4.6
10B	do	.96	.48	5.3	54	24.1	11.0	5.2	4.0	5.0
11B	do	.91	.48	5.2	60	24.5	15.0	N.D.	4.1	4.9
12B	do	.96	.48	5.6	58	29.4	15.0	?	4.0	4.9
13B	Corn-mold from 12B	.85	.44	5.3	60	19.8	6.7	2.0	6.6	4.5
14B	do	.98	.49	5.1	57	15.2	5.3	?	3.3	4.8
15B	Corn-mold from 13B	1.50	.60	5.0	48	28.4	10.0	2.3	2.8	5.0
16B	Corn-mold from 14B	.96	.48	5.2	60	29.4	15.0	4.6	5.1	4.6
17B	Corn-mold from 16B	.96	.48	5.0	54	32.1	19.2	4.0	2.6	5.0
18B	do	.96	.48	...	68	22.0	8.6	2.1	3.2	5.2
21B	Part corn-malt, part corn-mold	1.20	.60	5.1	72	21.4	8.9	Less than 2	3.5	4.9
22B	Corn-malt	.96	.48	5.5	56	34.8	11.0	4.7	2.7	4.9
23B	do [3]	...	...	...	...	...	...	...	...	...
25B	do	1.33	.44	5.6	52	39.3	13.8	N.D.	1.5	5.2
26B	do	1.15	.46	5.2	54	38.5	19.2	3.3	1.5	5.1
27B	do	1.11	.44	5.9	84	19.8	10.0	Less than 2	.25	6.7

1. All media contained about 0.02 percent ammonium bifluoride to inhibit growth of bacteria if a liquor became contaminated during run.
2. Milliliters of 0.1 N NaOH to titrate 10 ml. of filtered mold liquor to phenolphthalein end point.
3. Discarded after 24 hours because it was badly contaminated. Trouble traced to laboratory culture.

available was from malt beer. If a plant were to use fungal amylase regularly, the stillage would be from mold beer. It seemed conceivable that malt stillage might contain nutrients for mold growth and that these factors might not be present in mold stillage. In order to determine the effect of recycling mold stillage on the quality of the fungal amylase liquor, experiments 13B, 14B, 15B, 16B, 17B, and 18B were conducted in which mold stillage from a previous distillery operation was used for preparation of the medium. It was found that the use of recycled stillage from mold beers had no effect on the enzymatic potency of the resultant mold liquor or on the conversion power of the material when it was used in the distillery. These results are in accord with information obtained in pilot-plant work at the Northern Regional Research Laboratory *(5)*.

The principal difficulty in the preparation of the fungal amylase liquor was contamination of the medium during fermentation. In only three experiments was the finished liquor free of contaminants. A thorough search was made for the source of the infection; filtered air was tested for sterility, the cooked medium was sampled aseptically and cultured to determine the presence or absence of contaminating organisms, and the seed culture was examined. As a result of these tests it was decided that the large fungal amylase fermentor was the cause of the trouble. Because of its construction, this tank could not be sterilized with steam under pressure. In addition, the interior surface of the vessel was rough and difficult to clean.

The enzyme potency of the fungal amylase liquors varied widely. Saccharification powers as low as 15.2 percent in experiments 14B and as high as 39.3 in run 25B were obtained. It is believed that contamination was the chief cause of the variation. In pilot-plant work at the Northern Laboratory consistent results were obtained, but contamination was infrequent.

Data on the utilization of fungal amylase liquor in the distillery are given in Table 10-2. After a few runs in the distillery, it became evident that the power of the mold liquor to liquefy mash was less than that of malt. This is of special significance to a plant in which continuous cooking and conversion are employed, as only about 6 minutes are provided for liquefaction and primary conversion of the mash. If the mash is too viscous at the time it enters the cooler, the pressure drop across the cooler becomes excessive and the mash pump delivers less material. Thus, the over-all rate of mashing must be decreased in order to maintain steady-state conditions in the continuous systems.

When malt-converted mash was pumped at the rate of 400 bushels of grain per hour, the pressure at the discharge side of the mash pump was about 30 p.s.i.g. At the same rate, but with mash converted with fungal amylase, the pump pressure in most cases was 40 to 60 p.s.i.g. In one experiment, this pressure reached 100 p.s.i.g., which was sufficient to lower the capacity of the pump and mash accumulated in the vacuum flash cooler.

In order to improve liquefaction when fungal amylase was used for conversion, a small amount of malt was mixed with the mold liquor in most of the experiments. From 1 to 2 percent malt was employed on the basis of the total grain bill.

It was observed that when mash was converted with fungal amylase at a temperature of 152° F, rather than 145° to 146° F, liquefaction was improved. The higher temperature had no detrimental effect on the alcohol yields. It is believed that fungal amylase alone will liquefy mash satisfactorily if a conversion temperature of 152° to 155° F and a retention time of 12 to 15 minutes are used. Indeed, recent plant-scale experiments have shown this to be true. Another factor which affects rate of liquefaction is the quality of the fungal amylase liquor.

Experiment 26B was a very successful run. Corn mash for the second fermentor of this experiment was converted with fungal amylase mixed with only a very small quantity of malt (0.3 percent of the total grain bill); yet the mashing operation was normal in all respects. It is significant that the mold liquor was of good quality. It is reasonable to assume that the use of fungal amylase of consistently good quality will help to eliminate the liquefaction problem in these runs.

On the average, about 3.5 gallons of mold liquor was used for the conversion of 1 bushel of corn, as received. The controlling factor in these experiments was liquefaction during mashing operations. No attempt was made to determine the effect on fermentation of smaller amounts of liquor. However, in pilot-plant work and in recent plant-scale experiments satisfactory results were obtained with 2.7 gallons and less of fungal amylase liquor per bushel of grain when the liquor was of good quality and under favorable conditions for primary conversion.

Secondary conversion and fermentation of mold-converted mashes appeared to proceed more slowly than with malt-converted material. This indicates that a longer fermentation time might be required when fungal amylase is used unless more favorable primary conversion were obtained.

Table 10-2. Data on the Utilization of Fungal Amylase Liquor for the Conversion of Corn Mashes in the Alcohol Plant of the Grain Processing Corporation.

Converting Agent		Grain Mashed		Mashing Operations			
Mold Liquor Produced in Run Number—	Saccharification Value, Conversion	Total Corn Mashed For Two Fermentors, as Received	Grade of Corn Used[1]	Fermentor Number	Quantity of Mold Liquor Used Per Corn as Received	Malt Added to Help Thin Mash (Percent of Grain Bill)	Conversion Temperature
	Percent	Bushels			Gallons	Percent	°F
9B....	24.0	5,502	No. 3	9	4.0	0.0	145
				2	3.4	1.0	145
10B....	24.1	6,003	..do...	1	2.8	1.2	151
				3	3.8	1.0	151
11B....	24.5	6,130	..do...	3	3.2	.9	152
				5	3.9	.1	152
12B....	29.4	5.175	..do...	2	4.3	.0	152
				4	4.5	.0	152
13B....	19.8	6.413	..do...	2	3.2	1.0	153
				4	3.2	2.0	153
14B....	15.2	6,404	..do...	5	3.3	1.5	152
				7	3.4	1.4	152
15B....	28.4	6.473	..do...	2	3.1	1.3	151
				4	3.1	1.1	151
16B	29.4	6.030	..do...	3	3.3	2.0	153
				5	3.3	1.8	153
17B....	32.1	6,073	Sample, damage	8	3.5	1.2	153
23B...	See 119		50%	3	3.5	1.2	153
18B....	22.0	(4)	(4)	(4)	(4)	(4)	(4)
21B....	21.4	5,437	No. 3	4	2.2	2.0	150
				6	3.8	2.4	150
22B....	34.8	5,596	Sample, damage	8	3.6	1.2	152
			50%	3	3.9	1.3	152
		Table 1	...	...	...	...	...
25B...	39.3	7,007	No. 3	2	3.3	1.4	151
				4	3.3	1.3	151
26B...	38.5	6.961	..do...	6	3.2	1.1	153
				8	3.9	.3	153
27B....	19.8	6,999	..do...	5	3.4	4.6	145
				7	3.9	4.6	145

[1] Grain was graded No. 3 because of moisture content.

[2] Milliliters of 0.1 N NaOH to titrate 10 ml. of filtered beer to phenolphthalein end point.

[3] This number was obtained by dividing total volume of finished beer in fermentor, in gallons, by total bushels (56 pounds) of dry grain with which fermentor was charged.

[4] Mold liquor contaminated badly; not used in alcoholic fermentation.

	Fermentation						
			Acidity				
Type Stillage for Backset	Age of Beer When Fermentor Emptied	Final Specific Gravity	Titratable[2]	Final	Alcohol in Finished Beer	Final Concentration Grain in Beer Per Bu. (Dry)	Yield of Alcohol Per 56 lb. Dry Grain
	Hours	°Balling		pH	Percent	Gallons[3]	Proof gallons
Corn-malt	62	0.9	3.2	4.5	5.76	46.4	5.35
...do.....	59	.8	4.3	4.9	8.10	37.1	6.01
...do....	105	1.3	6.7	4.6	8.11	36.8	5.97
...do.....	104	1.4	5.8	4.6	8.02	37.1	5.95
...do...	117	.8	4.7	4.7	7.81	38.1	5.95
...do...	118	.8	5.6	4.6	7.58	40.6	6.16
...do...	107	1.3	6.6	4.6	7.55	41.3	6.24
...do...	105	1.1	4.5	4.7	7.94	39.4	6.26
Corn-mold	109	2.0	9.4	4.0	6.48	38.5	4.99
...do...	110	.9	6.1	4.8	7.89	37.0	5.84
...do...	54	2.5	8.4	4.9	7.14	36.4	5.20
...do...	49	3.1	7.9	5.0	6.32	36.7	4.64
...do...	110	1.5	8.0	4.2	7.24	38.4	5.56
...do...	110	.7	5.8	4.7	8.31	35.3	5.87
...do...	105	.7	5.4	4.4	7.45	39.6	5.90
...do...	93	.6	4.2	4.8	8.38	37.1	6.22
...do...	118	.9	5.0	4.9	8.78	35.3	6.20
...do...	114	1.0	7.8	4.1	7.87	36.6	5.76
(4)	(4)	(4)	(4)	(4)	(4)	(4)	(4)
Corn-malt	148	.3	5.7	4.1	7.44	36.9	5.49
...do...	142	.6	6.5	4.5	7.61	37.0	5.63
...do...	110	.2	6.5	4.5	8.12	37.8	6.14
...do...	96	.2	6.5	4.5	6.89	44.0	6.06
...	...	...	...	...	...	...	...
None	106	.0	4.1	4.6	7.61	39.4	6.00
...do...	105	.1	4.6	4.9	7.81	37.4	5.84
...do...	104	.1	2.1	4.8	7.76	39.0	6.05
...do...	104	.2	4.0	4.6	7.91	38.3	6.07
...do...	59	.2	3.2	4.6	7.94	37.8	6.00
...do...	52	1.7	4.5	4.6	7.93	36.6	5.79

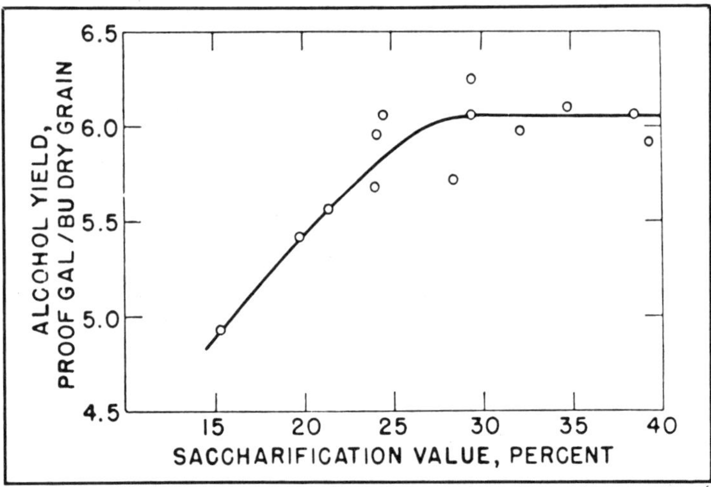

Fig. 10-3. The relationship between saccharogenic power of fungal amylase and yield of alcohol.

Samples of mold-converted mash taken at various times during the alcoholic fermentation were examined for bacterial contamination. The extent of contamination was less than usually encountered in malt-converted mashes. This was to be expected, as many bacteria are added to the mash with the malt.

The results of the plant experiments indicate a relationship between saccharogenic power of the fungal amylase liquor and the yield of alcohol obtained from fermented mashes converted with this agent. The correlation is not precise for two reasons: The quantity of mold liquor used for the conversion of 1 bushel of corn was not the same in all experiments; and a small amount of malt was employed with the fungal amylase in most of the runs. Figure 10-3 shows the superficial relationship between saccharogenic power of fungal amylase and the yield of alcohol. For the preparation of the graph, the yield used was the average of those obtained in the two alcoholic fermentations conducted with a given batch of fungal amylase liquor. The data from plant run 27B were not used because of the inordinately large quantity of malt that was employed with the fungal amylase for conversion. It is indicated from the graph that to be a satisfactory converting agent the fungal amylase liquor should have a saccharification value of 25 percent or more. Of the 16 runs reported in Table 10-1, 10 batches of mold liquor had a saccharification value of 24 percent or more. The other 6 preparations were poor because of contamination of the culture during the mold fermentation.

Table 10-3. Properties of Products Made With Fungal Amylase as Converting Agent.

Run number	Alcohol: Chemical analysis			Organoleptic evaluation [1]		Feeds: Dried grains				Feeds: Solubles			
	Esters	Aldehydes	Fusel oil	Odor	Taste	Moisture	Protein	Fat	Fiber	Moisture	Protein	Fat	Fiber
	G./100 l	G./100 l	G./100 l			Percent	Percent	Percent	Percent	Percent	Percent	Percent	Percent
9B	1.208	0.386	1.9	Good	Good	7.4	23.1	10.1	9.2	5.6	32.6	13.4	4.2
10B	.880	.385	2.3	do	do	8.0	24.1	10.2	9.2	4.3	26.7	12.1	3.8
11B	1.408	.375	2.1	Rejected	do	6.8	22.5	9.3	9.0	5.6	34.2	14.3	5.1
12B	.440	.380	2.3	Good	do	9.0	27.1	10.6	8.9	4.8	30.4	14.1	5.0
13B	1.320	.386	1.9	do	do	6.7	24.3	10.0	10.5	5.4	28.4	12.9	4.6
14B	2.476	.374	2.0	do	do	8.2	26.4	9.6	10.3	5.2	29.6	12.6	4.7
15B	1.584	.440	2.1	Rejected	Rejected	7.9	22.4	10.0	9.4	5.1	33.8	12.8	4.5
16B	.886	.397	2.2	Good	Good	8.6	24.1	10.2	8.8	4.6	27.6	14.2	4.8
17B	.528	.385	2.3	do	do	6.7	24.5	10.8	10.3	5.6	33.7	14.6	4.5
21B	2.204	.378	2.1	do	do	9.4	22.8	9.8	9.4	4.3	34.6	14.2	3.9
22B	1.606	.386	2.1	do	do	9.6	24.6	10.1	9.8	5.4	28.6	13.7	3.6
25B	2.486	.385	2.2	do	do	8.7	26.5	10.1	10.2	5.2	30.3	12.4	4.1
26B	.896	.392	2.1	do	do	9.1	23.4	10.4	9.4	4.9	31.0	14.5	3.8
27B	1.488	.384	2.3	do	do	6.4	25.1	9.6	9.2	5.1	28.7	13.7	4.2
Average	1.386	.388	2.14			8.0	24.4	10.1	9.5	5.1	30.7	13.5	4.3
Typical values [2]	1.220	0.392	2.20			9.4	24.9	9.5	9.9	6.0	32.0	12.5	4.5

[1] Evaluations relative to high-grade spirits made with malt. Frequency of rejection of mold alcohol no greater than that of spirits made with malt.
[2] Typical values for products made with malt as converting agent.

In the 10 plant runs in which fungal amylase of satisfactory quality was used, the average alcohol yield was 5.98 proof gallons per bushel of dry grain. During this same period of operation, the average yield of alcohol from malt-converted mashes was 5.87 proof gallons per bushel of dry grain. In experiments 9B, 12B, and 26B, four fermentors were set with mash converted with fungal amylase which had been supplemented with only a trace of malt, or in which no malt was used. It is significant that the average alcohol yield from these fermentors was 5.98 proof gallons per bushel of dry grain. This indicates that the 1 to 2 percent malt used in other runs to help thin the mash did not affect the alcohol yields.

Heat-Damaged Corn in Experiments

In plant runs 17B and 22B, corn that was approximately 50 percent heat-damaged was mashed to determine whether fungal amylase could be used with this type of material. The results were satisfactory. An average alcohol yield of 6.04 proof gallons per bushel of dry grain was obtained from the four fermentors of these runs. The average yield from this type of corn converted with malt was 5.86 in fermentations conducted immediately before and after each of the experimental runs.

No difficulties or differences in operation were encountered in the distillation of beer from mold-converted mash. Feed recovery operations also were normal. Feed yields, although impossible to measure accurately, appeared to be the same as yields from malt stillage.

The use of fungal amylase for the conversion of yeast mashes was not investigated. All yeast cultures used in the plant experiments were prepared with malt-converted mash.

Quality of Alcohol and Feeds Produced with Fungal Amylase

Samples of the alcohol and byproduct feeds recovered during the experimental runs were analyzed chemically for components which are used commonly as a means of evaluation. In addition, the alcohol was compared, organoleptically, with standard samples of high-grade spirits made with malt. The results of these tests are given in Table 10-3. Included in this table is a typical analysis of products made with malt as the converting agent. There is no significant difference between these figures and the average values for the products made with fungal amylase as the saccharifying agent. The organoleptic evaluation of the alcohol produced from mold-converted mash showed it to be satisfactory.

Quantities of distillers' dried grains and distillers' dried solubles produced by the fungal amylase process and by malt conversion were supplied to the Experiment Station of the University of Nebraska and to the Bureau of Animal Industry, U. S. Department of Agriculture, for comparison by animal feeding tests. Preliminary reports have been made by each of these agencies on the progress of their experiments. [5, 6, 7] At the Nebraska Experiment Station steer calves which were fed grain and prairie hay along with the fungal amylase grains made an average daily gain of 1.8 pounds for 223 days, and required an average of 286 pounds of prairie hay, 599 pounds of ground shelled corn, and 146 pounds of distillers' product per 100 pounds of live-weight gain. The lot of calves fed the dried grains from malt-converted mash made an average daily gain of 1.68 pounds, and required an average of 302 pounds of prairie hay, 602 pounds of ground shelled corn, and 155 pounds of dried grains per 100 pounds of live-weight gain. A similar lot of cattle was fed prairie hay and ground shelled corn without any protein supplement. These animals made an average daily gain of 1.35 pounds and required 441 pounds of hay and 838 pounds of ground shelled corn for 100 pounds of gain.

In another series of tests by the Nebraska Experiment Station steer calves were wintered on a full feed of August-cut prairie hay, 2 pounds per head daily of ground shelled corn, and 1.72 pounds per head daily of the distillers' products. The lot fed fungal amylase grains made an average daily gain of 1.21 pounds and consumed an average of 880 pounds of hay, 165 pounds of corn, and 144 pounds of distillers' grains per 100 pounds of gain. The lot of calves fed grains from malt-converted mash made an average daily gain of 1.19 pounds and required an average of 876 pounds of hay, 168 pounds of corn, and 146 pounds of distillers' product per 100 pounds of live-weight gain.

The results of the Nebraska tests indicate that there is no significant difference in feed value between distillers' grains made with fungal amylase or with malt as the converting agent.

Investigations of the following type have been conducted or are in progress by the Bureau of Animal Industry on the byproduct feeds made with malt and with fungal amylase: A complete analytical study, including the amino acid and vitamin content of the

---

[5] Baker, M. L., Col. of Agr. and Agr. Expt. Sta., Univ. of Nebr. (Private communication.)

[6] U.S. Dept. Agr., Bur. of Anim. Indus. Prog. Rept. (unpublished.)

[7] Dowe, T. W., and Arthaud, V. H., Col. of Agr. and Agr. Expt. Sta., Univ. of Nebr. Prog. Repts., No. 192 and No. 193. (Unpublished.)

byproducts produced by the two processes; toxicity and palatability studies with sheep and cattle; poultry- and swine-feeding tests where the solubles are used as vitamin carriers; and digestibility studies and practical sheep-feeding tests in which the dark grains are used as protein supplements.

The Bureau of Animal Industry found no significant difference in either chemical composition or amino acid content between comparable materials produced with malt and fungal amylase. The results to date of the palatability and toxicity tests with sheep indicate that the fungal amylase dark grains had no toxic or detrimental effect on the animals, although they were fed at four to seven times the level ordinarily used in feeding work and for long periods of time. There are some indications that sheep have a slight preference for the conventional malt dark grains over the fungal amylase byproducts; however, the dark grains produced by the fungal amylase process were palatable, as they were consumed in large quantities throughout the test periods.

Practical sheep-feeding tests have shown that distillers' dark grains produced by either the malt or fungal amylase process can satisfactorily replace linseed oil meal as a protein supplement in a fattening ration for lambs.

No significant differences have been observed between solubles produced by either process when used in swine rations.

Preliminary chick feeding tests have been conducted by the Bureau of Animal Industry, in which the conventional malt solubles and also the fungal amylase solubles have been used as riboflavin carriers.

A large portion of this report concerned the estimated costs for installation and operation of a fungal amylase plant. Given the fact that the report was written several years ago and today's current rate of inflation, that section of the report could have been written in rubles for all the good it would do us. Consequently, I have not included it.

However, there is a bibliography you might find useful. If you want any of the following articles, simply ask your local librarian to get them for you on inter-library loan.

ADDITIONAL READING

(1) Erb, N. M., and Hildebrant, F. M.
1946. *Mold as an Adjunct to Malt in Grain Fermentation.* Ind. and Eng. Chem. 38:792-794.

(2)——, Wisthoff, R. T., and Jacobs, W. L.
1948. *Factors Affecting the Production of Amylase by Aspergillus Niger, Strain NRRL 337, When Grown in Submerged Culture*. Jour. Bact. 55:813-821.

(3) Grove, Otto.
1914. *The Amylo Process of Fermentation*. Jour. Inst. of Brewing. 20:248-266.

(4) Le Mense, E. H., Corman, Julian, Van Lanen, J. M., and Langlykke, A. F.
1947. *Production of Mold Amylases in Submerged Culture*. Jour. Bact. 54:149-159.

(5)——, Sohns, V. E., Corman, Julian, Blom, R. H., Van Lanen, J. M., and Langlykke, A. F.
1949. *Grain Alcohol Fermentations: Submerged Mold Amylase as a Saccharifying Agent*. Ind. and Eng. Chem. 41:100-103.

(6) Sandstedt, R. M., Kneen, Eric, and Blish, M. J.
1939. *A Standardized Wohlgemuth Procedure for Alpha-Amylase Activity*. Cereal Chem. 16:712-729.

(7) Somogyi, Michael.
1945. *A New Reagent for the Determination of Sugars*. Jour. Biol. Chem. 160:61-68.

(8) Tsuchiya, Henry M., Corman, Julian, and Koepsell, Harold J.
1949. *The Production of Fungal Amylase in Submerged Culture*. Cereal Chem. 27:322-330.

(9) Underkofler, L. A., Fulmer, Ellis I., and Schoene, Lorin.
1939. *Saccharification of Starchy Grain Mashes for The Alcoholic Fermentation Industry: Use of Mold Amylase*. Ind. and Eng. Chem. 31:734-738.

(10)——, Severson, G. M., and Goering, K. J.
1946. *Saccharification of Grain Mashes for Alcoholic Fermentation: Plant-Scale Use of Mold Amylase*. Ind. and Eng. Chem. 38:980-985.

(11) Van Lanen, J. M., and Le Mense, E. H.
1946. *The Production of Fungal Amylases in Submerged Culture and Their Use in the Production of Industrial Alcohol*. (Abstract.) Jour. Bact. 51:595.

(12) Woolner, Adolph, Jr., and Lassloffy, Aladar.
1909. *U. S. Patent 923,232* (June 1, 1909). *Manufacture Alcohol*. Chem. Abst. vol. 3, pt. 2, p. 2198.

If you're really heavy into lab techniques, there is also a good section on troubleshooting. If terms like *ml* for milliliter confuse you just go to a chemical supply house and ask for a 10-ml graduated cylinder. You can simply pour in the liquid and see what you get—up to 10 ml. of course. A "ml." is $\frac{1}{1000}$ th of a liter. That is not quite as much as a quart.

Turbidity simply means a cloudiness. It is something like what you get when you spill a drop of milk into a glass of water.

Mg is a milligram. A thousand "mg" makes up a gram. A thousand grams makes up a kilo. One kilo is 2.2 pounds. 0.01 N means $\frac{1}{100}$ th of a mole. A *mole* is 6.02×10^{23} atoms or molecules of whatever you are dealing with per liter of solution.

Things like *HCl* mean hydrogen and chlorine combined, such as—in this case—hydrochloric acid. The capital letters stand for elements and the small letters simply enable you to tell atoms of different species apart. *Cl* for example is chlorine while *C* is carbon. If you want to find out what comprises a molecule from a formula—such as KI—just add things up on the periodic table. *KI* is potassium iodide. *K* for potassium—a metal—and *I* for iodide. Armed with that information, see if you can translate the following.

DETECTION OF BACTERIAL CONTAMINANTS IN FUNGAL AMYLASE MEDIUM AND IN FILTERED AIR

A sample of mold liquor is withdrawn aseptically from the fermentor or seed tank into a sterile flask. One ml. of this is transferred with a sterile pipette to a large test tube that contains 10 ml. of nutrient broth. The test medium is prepared in accordance with the directions given below. The broth culture is incubated for 24 hours at 88°F. A loopful of this culture is then transferred to a second tube of broth which is prepared just like the first one. The second tube is incubated for 24 hours at 88°F and then examined for turbidity. If the broth is turbid, bacterial contaminants are present. The result should be confirmed by examination of the broth under the microscope. See Table 10-4.

Sterility of air used in the fermentor and seed tank is tested by the passage of a sample stream of it through 200 ml. of the described nutrient broth for 30 minutes. The broth is contained in a 1,000-ml. Erlenmeyer flask equipped with an air delivery tube, and with an opening plugged with cotton through which the air is vented. After aeration of the broth it is incubated for 24 hours at 88°F. Sterility is shown by the absence of turbidity in the medium. The result should be confirmed by examination of the broth under the microscope.

DETERMINATION OF ALPHA-AMYLASE ACTIVITY IN FUNGAL AMYLASE PREPARATIONS

The methods for the determination of alpha-amylase activity in fungal amylase preparations are based on the procedure of Sandstedt, Kneen, and Blish (6) and that given in the Journal of the Association of Official Agricultural Chemists, fifth edition (1947), page 96. The method depends on the length of time required for an enzyme preparation to hydrolyze beta-amylase treated starch to dextrins which give a characteristic color when the enzyme-substrate mixture is added to a solution of iodine and potassium iodide.

Table 10-4. Directions for Preparing Test Medium.[1]

Nutrient broth		Salt solution	
Glucose	1 gm.	$MgSO_4 \cdot 7H_2O$	4 gm.
Yeast extract	1 gm.	$FeSO_4 \cdot 7H_2O$	0.2 gm.
Peptone	1 gm.	$MnSO_4 \cdot 4H_2O$	0.8 gm.
K_2HPO_4	0.5 gm.	NaCl	0.2 gm.
Sodium citrate	1 gm.	Conc. HCl	0.4 ml.
Sodium acetate	0.1 gm.	Water to make	100.0 ml.
Salt solution	2.0 ml.		
Water to make	100.0 ml.		

[1] Dispense 10-ml. quantities into test tubes; plug with cotton; sterilize 15 minutes at 250° F.; and cool.

Reagent

A *reagent* is merely a chemical that reacts with or indicates that another substance is present.

(a) Stock iodine solution: 5.5 grams of iodine crystals and 11 grams of potassium iodide are dissolved in water and the solution is made up to 250 ml. The stock solution is stored in the dark. A fresh solution is made each month.

(b) Dilute iodine solution: 2 ml. of the stock iodine solution (I*a*) is added to 20 grams of KI dissolved in water and made up to 500 ml.

(c) Buffer solution: 120 ml. of glacial acetic acid and 164 grams of anhydrous sodium acetate are dissolved and made up to 1 liter.

(d) Beta-amylase: A special beta-amylase in dry form has been developed by Wallerstein Laboratories, New York, N.Y., for use in the preparation of the alpha-amylodextrin substrate in the determination of alpha-amylase in starch hydrolyzing materials by dextrinization procedures. The beta-amylase is standardized at 2,000° Lintner and to the specifications set by the Malt Evaluation Committee of the American Association of Cereal Chemists. The specifications require that "(1) at the addition level recommended there shall be a variation not greater than 5 percent in the dextrinization times of a given malt extract when one- and three-day old substrates are compared, and (2) a substrate prepared by adding twice the recommended level of beta-amylase shall not deviate by more than 5 percent from one prepared with the recommended level after 24 hours standing."

(e) Buffered beta-amylase limit dextrin (alpha-amylodextrin) solution: A suspension of 10 grams (dry weight) of soluble starch (according to Lintner, "Special for Diastatic Power Determination") is poured slowly into boiling water. The solution is boiled for 1 to 2 minutes with stirring and then cooled. 25 ml. of buffer solution and 250 mg. of beta-amylase (I*d*) dissolved in a small amount of water are added to the starch solution. The volume is made up to 500 ml. and saturated with toluol. The solution is stored at, or close to, 30°C for not less than 24 nor more than 72 hours before use.

(f) Standard solution: 100 ml. of 0.01 *N* HCl are added to 25 grams of $CoCl_2 \cdot 6H_2O$ and 3.84 grams of $K_2Cr_2O_7$. 10-ml. portions are placed in test tubes which are then sealed. These standard solutions may be kept indefinitely.

Enzyme Reaction

(1) The fungal amylase preparation (culture filtrate), the substrate (I*e*) and iodine (I*b*) solutions are all attempered to 30°C. Five ml. of fungal amylase preparation (or appropriate dilutions thereof) is added to 10 ml. of the substrate solution (I*e*). The instant that the fungal amylase preparation is added to the substrate solution, the time is noted. At appropriate time intervals, 1-ml. samples of the reaction mixture are transferred to test tubes containing 5 ml. of the iodine solution (I*b*). The optical density of the mixture is compared with that of the standard (I*f*) in a rapidly acting photometer (type similar to Lumetron model No. 400 with a 650μ filter). The instrument is set so that the standard (I*f*) gives 50 percent transmission. When the reaction has proceeded to the

point where the enzyme-substrate solution gives 50 percent transmission when added to the iodine solution (I*b*) the end point has been reached and the time is noted. The elapsed time is referred to as the dextrinization time.

Samples should be taken and read every 15 seconds as the end point is approached. Although a photometer is desirable and convenient to use in this determination, visual reading may be made when such an instrument is not available.

Calculation

The alpha-amylase activity can be calculated from the following equation, where D.T. is the dextrinization time.

$$\frac{0.2}{1} \times \frac{60}{\text{D.T.}} \times \frac{\text{diln.}}{\text{ml. enz. sol.}} = \text{grams beta-amylase treated starch dextrinized per ml. of original enzyme solution per hour.}$$

Table 10-5. Grams of Beta-Amylase Treated Starch Dextrinized per ml. of Original Enzyme Solution per Hour.

Minutes	0	¼	½	¾
10	0.240	0.234	0.229	0.223
11	.218	.213	.209	.204
12	.200	.196	.192	.188
13	.185	.181	.178	.175
14	.171	.168	.166	.163
15	.160	.157	.155	.152
16	.150	.148	.145	.143
17	.141	.139	.137	.135
18	.133	.132	.130	.128
19	.126	.125	.123	.122
20	.120	.119	.117	.116
21	0.114	0.113	0.112	0.110
22	.109	.108	.107	.105
23	.104	.103	.102	.101
24	.100	.099	.098	.097
25	.096	.095	.094	.093
26	.092	.091	.091	.090
27	.089	.088	.087	.086
28	.086	.085	.084	.083
29	.083	.082	.081	.081
30	.080			

Example No. 1:

5 m. of 1:100 diln. of fungal amylase has a D.T. of 10 minutes.

$$\frac{0.2}{1} \times \frac{60}{10} \times \frac{100}{5} =$$ 24 grams beta-amylase treated starch dextrinized per ml. of original enzyme solution per hour.

Example No. 2:

1 ml. of 1:20 diln. of fungal amylase plus 4 ml. of water has a D.T of 10 minutes.

$$\frac{0.2}{1} \times \frac{60}{10} \times \frac{20}{1} =$$ 24 grams beta-amylase treated starch dextrinized per ml. of original enzyme solution per hour.

For convenience, Table 10-5 has been compiled from the foregoing equation.

The use of the table is demonstrated in the following examples:

5 ml. of 1:100 diln. of fungal amylase has a D.T. of 10 minutes.

10 min. = 0.24; diln. is 1:100.

(0.24) (100) = 24 grams of beta-amylase treated starch dextrinized per ml. of original enzyme solution per hour.

1 ml. of 1:20 diln. of fungal amylase plus 4 ml. of water has a D.T. of 10 minutes.

10 min. = 0.24; diln. is 1:20 and 1:5.

(0.24) (20) (5) = 24 grams of beta-amylase treated starch dextrinized per ml. of original enzyme solution per hour.

Precautions

(1) The type of starch used is very important. Soluble starch prepared according to Lintner, "Special for Diastatic Power Determination," must be used. Emphasis is placed on the term "Special for Diastatic Power Determination." The starch must be of grade equivalent to Merck's soluble starch, according to Lintner, "Special for Diastatic Power Determination."

(2) In the preparation of buffered beta-amylase limit dextrin (I*e*), the starch must be in complete solution after boiling for 1 to 2 minutes. Formation of starch film must be avoided by pouring while hot into the volumetric flask and cooling quickly prior to the addition of the buffer (I*c*) and water to make up to volume.

DETERMINATION OF MALTASE ACTIVITY IN FUNGAL AMYLASE PREPARATIONS

The method, developed at the Northern Laboratory for the determination of maltase activity in fungal amylase preparations, is based on the observation that an increase of 78 percent in reducing power is obtained when maltose monohydrate is hydrolyzed to glucose under the conditions given below. There is a stoichiometric relationship between the amount of enzyme and the hydrolysis rate, when the rate is calculated from the difference in maltose hydrolyzed at 15 and 120 minutes.

Watch this word *reducing* in chemistry. It means *gaining electrons. Oxidizing* means *losing electrons.* In the early days of chemistry, somebody decided that electricity flowed in exactly the opposite direction of what it actually does.

Electricity flows from the "negative" pole of a battery simply because that's how it was originally mislabeled. All of chemical literature has had to be written (and read) backwards ever since.

Back to the directions.

Reagents

(*a*) pH 4.4 acetate buffer (6.0*M*), 217 ml. of glacial acetic acid, and 183 grams of anhydrous sodium acetate are dissolved and diluted to 1 liter with water.

(*b*) Acetate buffer (0.3*M*), maltose substrate (0.06*M*) solution. 2.35 grams of maltose monohydrate (92 percent pure as calculated on reducing value, for example, of Eastman Kodak Co. product) and 5 ml. of acetate buffer are diluted to 100 ml. with water.

(*c*) 1*N* sulfuric acid, 1*N* sodium hydroxide, phenolphthalein indicator.

(*d*) Reagents for sugar estimation by method of Somogyi (*7*).

Enzyme Reaction

Five ml. of fungal amylase preparation (culture filtrate) and 10 ml. of buffered substrate solution, both attempered to 30°C., are placed in a test tube and the tube is incubated in a water bath at 30°C. After 15 minutes, a 3-ml. aliquot of the reaction mixture is transferred to a 100-ml. volumetric flask containing 3 ml. of 1*N* H_2SO_4 to inactivate the enzyme. After 120 minutes, a second 3-ml. aliquot of the reaction mixture is treated in similar manner.

Analysis

After 10 minutes, the acidified reaction mixtures are adjusted to the phenolphthalein end point with 1*N* sodium hydroxide solution and made up to 100 ml. with water. Five ml. aliquots are taken for analyses for reducing value (R.V) by the method of Somogyi (*7*), using the 20-minute heating period.

In this procedure, the R.V.'s of the reaction mixtures, obtained after 15 and 120 minutes' hydrolysis, are measures of glucose produced, residual maltose, and reducing sugars in the enzyme preparation.

Calculation

a = R.V. of reaction mixture incubated for 15 minutes.

b = R.V. of reaction mixture incubated for 120 minutes.

$$\frac{(b-a)}{0.78} \times (\text{glucose equivalent of } Na_2S_2O_3 \times 1.78) \times 20 \times \frac{60}{105} = \text{mg.}$$

maltose hydrolyzed/ml. of enzyme preparation in 1 hr.

Precautions

Hydrolysis rate values should be between 2 and 10 mg. maltose hydrolyzed/ml. enzyme preparation in 1 hr. to be acceptable. Values in higher range are preferred.

A pH of 4.4 must prevail in reaction mixture. Fungal amylase preparations highly buffered at pH values other than 4.4 must be adjusted to approximately this point before testing.

DETERMINATION OF SACCHARO-GENIC ACTIVITY OF FUNGAL AMYLASE

The method of analysis was adapted by Erb, Wisthoff, and Jacobs (*2*) from a combination of two analytical procedures—methods 20.61, p. 257, and 20.28, pp. 244-245—given in the Journal of the Association of Agricultural Chemists, sixth edition (1945).

Reagents

(*a*) Starch solution (3 percent). Make a paste of 3 grams of Lintner soluble starch with cold water. Pour slowly into about 70 ml. of boiling water. Cool and add water to 100 ml.

(*b*) Buffer solution. Make 6 ml. of glacial acetic acid and 8.2 grams of anhydrous sodium acetate to 1 liter with water. The pH of this solution should be 4.6 to 4.8.

(*c*) Sulfuric acid solution. (3.58*N*). 50 ml. conc. H_2SO_4 diluted to 500 ml. with water.

(*d*) Sodium tungstate solution. 12 grams diluted to 100 ml. with water.

(*e*) Ferricyanide solution. (0.1 *N*). 33.0 grams of pure dry $K_3Fe(CN)_6$ 44.0 grams of anhydrous Na_2CO_3 per liter.

(*f*) Acetic acid-salt solution. Make up 200 ml. of glacial acetic acid, 70 grams of KCl, and 40 grams of $ZnSO_4 \cdot 7H_2O$ to 1 liter with water.

(*g*) Starch and KI solution. Add 2 grams of soluble starch to small quantity of cold water and pour slowly into boiling water with constant stirring. Cool thoroughly (or resulting mixture will be dark colored), add 50 grams of KI and make up to 100 ml. with water. Add 1 drop of NaOH solution (saturated).

(*h*) Thiosulfate solution. 24.82 grams of $Na_2S_2O_3 \cdot 5H_2O$ and 3.8 grams of $Na_2B_4O_7 \cdot 10H_2O$ per liter.

Procedure

The following are placed in a 100-ml. Erlenmeyer flask:

20 ml. starch solution (3-percent Lintner soluble starch solution).
25 ml. buffer solution.
1 ml. fungal amylase (filtrate from mold culture).

Hold for 1 hour at 30°C. in a water bath. Add 2 ml. of 3.58 *N* H_2SO_4 and 2 ml. of sodium tungstate solution. Filter at once; discard first 8 to 10 drops. Pipette a 5-ml. aliquot into a 1- by 8-inch Pyrex test tube and add 10 ml. ferricyanide solution. Hold exactly 20 minutes in a boiling water bath. Cool to room temperature and empty the tube into a 100-ml. Erlenmeyer flask. Rinse with 25 ml. acetic salt solution, add 1 ml. starch + KI solution, and titrate with 0.1 *N* thiosulfate solution. Record ml. of thiosulfate used. Run a blank containing only starch and buffer in each test.

Calculations

Use table on page 245, A.O.A.C., sixth edition (p.216, A.O.A.C., 5th edition), to find the amount of maltose corresponding to the 0.1 *N* ferricynade used. Divide by 20 to give the mg. of maltose per 60 mg. of starch. This figure divided by 60 gives the percentage of conversion in 1 hour at 30° C.

Example

Blank	9.68 ml.
Test	1.52 ml.
	8.16 ml. = 475 mg. maltose

$$\frac{475}{20} \times \frac{1}{60} = 39.58 \text{ percent conversion}$$

If the results show over 40 percent conversion, a smaller aliquot than 5 ml. may be taken for the ferricyanide test and the results calculated accordingly.

Go back to "Additional Reading" for such things as "somogyi (7)." For further assistance, call a local chemical or microbiology supply house. This chapter should give you enough to ask questions and to understand a few of the technical answers you'll get.

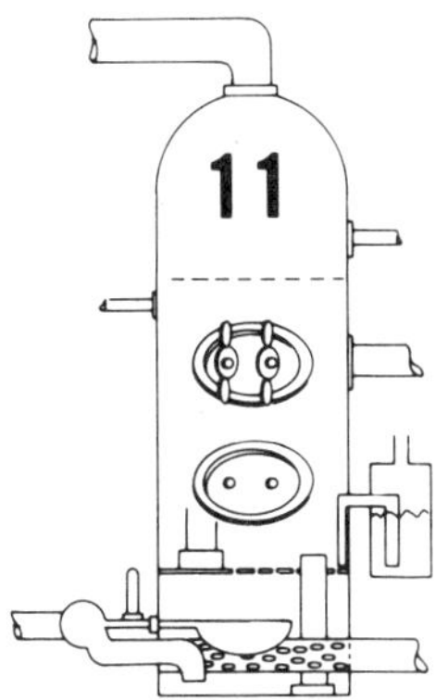

Continuous Fermentation

The production of alcohol from sugar crops has almost faded into oblivion in this country. Accurate information regarding "how it's done" other than on a backyard barbecue scale seems also to have faded. A lot of the literature is just plain wrong.

Alcohol from molasses is apparently still "big business" in the Soviet Union. Consequently, accurate information on the process is much easier to obtain. Let's hope my translation of Russian information is accurate. Certain idiomatic expressions in Russian simply do not translate into English. For example, apparently a "hot water" pipe in Russian can be a "steam" or a "hot water" pipe in English.

I have scaled the whole operation down to junkyard size in order to make the Russian methods affordable and accessible to the basement, backyard, or barnyard producer of motor fuel. The Russian idea of producing alcohol from molasses consists of beginning the procedure by bringing in railroad tank cars full of thousands of tons of the stuff. That's a little impractical for a fellow in downtown Des Moines, Iowa with only a basement in a three-bedroom house to work in.

METHODS

There are two ways to ferment molasses: batch, and continuous. The *batch method* consists of simply throwing the molasses in a big tank or 55-gallon drum, diluting it to 15 percent concentration molasses and 85 percent water. Warm it up between 65° and 70° and throw in a packet of baker's yeast. Let it ferment for two or three days before you distill it.

This is a relatively trouble-free method of doing things. If the mixture becomes contaminated by outside microorganisms and doesn't ferment properly, you can always throw it out, wash out the drum with a garden hose and start over.

It's a little more tricky with *continuous fermentation*. If you get hostile bacteria into the process, cleaning it out becomes a major undertaking.

BARNYARD OPERATION

A barnyard operation would proceed in the following fashion. This procedure can be used for almost any type of waste sugar such as leftover syrup from a canning factory. Before you follow the exact dilution factors for molasses, determine which other sugar-containing substance you are going to use. It's a whole lot easier to simply fill a series of coffee cans with different concentrations of fermentable products and water—pitch in a tiny but equal amount of yeast into each one—and see which one or ones ferment the most rapidly, than to look all over the chemical literature for the right concentrations for each particular product. You probably wouldn't find it anyway.

For a barnyard, you want an initial reservoir in which to dump the product. You might also be able to set the whole operation up in a large backyard, but the neighbors would probably have fits—especially when all the sticky stuff started attracting flies.

From the initial dumping reservoir, the molasses, peach syrup, or whatever (referred to as the *feedstock*), are pumped into storage tanks. Normally, there will be a filter between the initial dumping reservoir and pump to remove rocks, small birds, dead cats, or whatever fell in the feedstock before you got it. A screen or series of screens will do.

From here the feedstock has a water solution of calcium hypo-chloride dumped in it. A more common name for calcium hypo-chloride is household bleach. A mechanical mixer mixes the bleach and feedstocks together in a homogenous fashion. The reason for the addition of bleach is to kill any possible microflora that might eventually interfere with the desired fermentation processes. Let it alone for a day or two. The bleach—an antiseptic in the distillery process—must be given time to take hold.

MECHANICAL MIXER

Often a mechanical mixer can be installed in the middle of everything and the pumps direct everything to it. A schematic for such an arrangement is shown in Fig. 11-1. The mechanical mixer

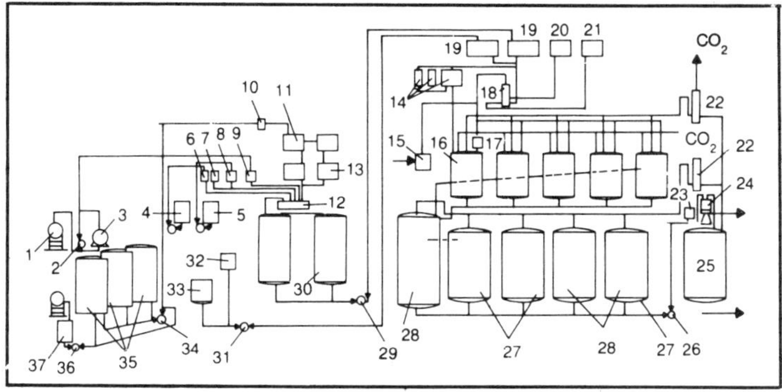

Fig. 11-1. Continuous fermentation.

is connected to almost everything that doesn't have wheels. The round gizmos on the accordian-looking stands are weighing stations. The small round jobs are hydraulic pumps. Everything else, except those items labeled 10 and 12, are storage containers. Item 10 is a testing station and 12 is the mechanical mixer. If you want to try to read the schematic, you have to do so backwards. You begin at number 37; that's the first step.

In Fig. 11-1, the antiseptic is added at point 30, the bleach is added into the mechanical mixer, 12, and then the resulting solution lands in 30. The bleach comes from measuring drum 6. You can probably substitute a series of 5-gallon buckets for stations 6 through 9. The other acids contained in those stations are as follows:

7 hydrochloric acid.
8 ortophosphoric acid.
9 sulfuric acid.

Number 4 is the bleaching powder stirring station. Number 5 is for nitrate and superphosphate, food for the yeast. For example, nitrate and superphosphate is pumped into the ortophosphoric acid and then into the mixer, which in turn pumps it with the feedstock into the two storage tanks, 30.

At this point, you could dispense with almost all the apparatus on the right of the schematic for a barnyard operation. I'll go ahead and list some of it anyway in case you might want to get fancy some day:

10 automatic sampler (quality of feedstock).
11, 37 receiving reservoirs.
13 weighing scales.
16 yeast generator.

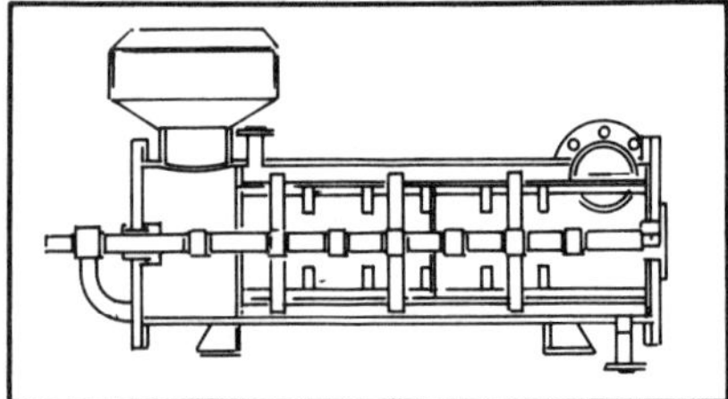

Fig. 11-2. A mechanical mixer.

17 automatic instruments for dispersement or dilution of syrup.

19 reservoirs, prior to processing.

22 alcohol traps.

23 filters.

24 yeast separators.

25 equivalent of beer well, last stop before distillation.

26 another pump or pumps.

27 fermentation station mixers.

28 fermentation stations.

31 usual hydraulic pump.

32 measuring reservoir.

33 molasses from cane receiver.

FERMENTATION STATIONS

The key to the whole barnyard system is the way the fermentation stations are set up. You could set up a yeast generator if you wanted, but it's hardly necessary. This schematic illustrates a factory that is also heavily engaged in the production of yeast for baking for sale on a large commercial scale.

If you decide to get fancy, you might want to duplicate the Russian mechanical mixer. A shaft turns in a cylinder onto which are attached three paddle wheels. A motor, not shown in the illustration, drives the shaft from the left-hand side. The curved job on the left is merely a brace with a bearing in it to support the rotating shaft. If you need to mount one of these over your storage tanks, you will probably want to run a chain or belt drive to an electric motor on the floor. If something goes wrong with your power supply, you won't have to crawl into the rafters to try to fix it or replace it.

The antiseptics and yeast food are dumped into the hopper on the top left (Fig. 11-2). After the paddle wheels have mixed everything, the solution flows out the round hole facing you on the top right. A drain valve is installed on the lower right of the

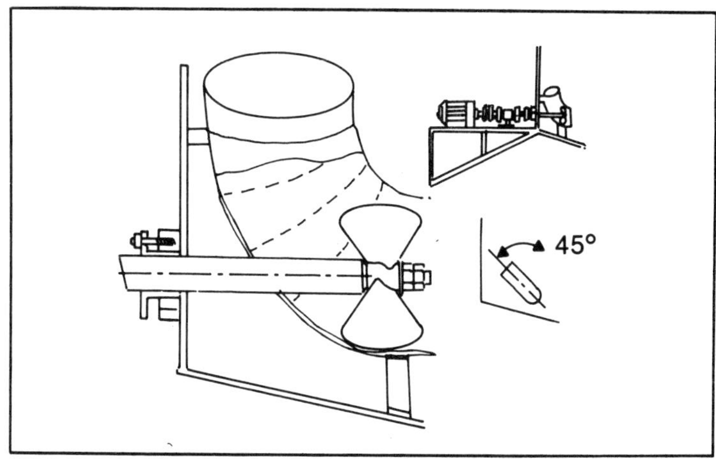

Fig. 11-3. An agitator propeller.

cylinder. Cleaning is accomplished by sticking a water hose in the hopper and unplugging the drain valve. If something really serious happens—like a piece of internal machinery breaking—the covers on each end are unbolted and the entire internal assembly is simply slid out for repair or replacement of the internal parts.

There are a few other pieces of equipment you might find interesting. One is an agitator propeller for a yeast tank. Fermentation proceeds best with no disturbance, but yeast propagation is most rapid if the mixture is kept agitated. If you look closely at Fig. 11-3, you will see that the curved pipe is actually inside the yeast tank and it is capped at the top end.

To keep them continuously operating, the yeast generator's are laid out in a circular pattern (Fig. 11-4). The layout of the yeast generator is noteworthy primarily because the fermentation stations are organized in much the same way. Just imagine that these things are 55-gallon drums or 250-gallon used propane tanks.

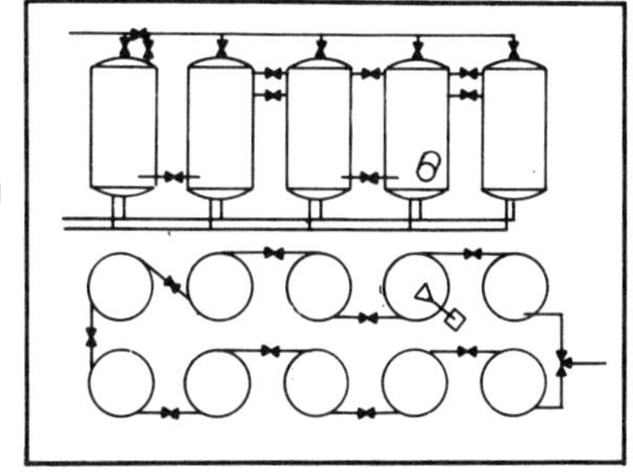

Fig. 11-4. A continuously operating yeast installation.

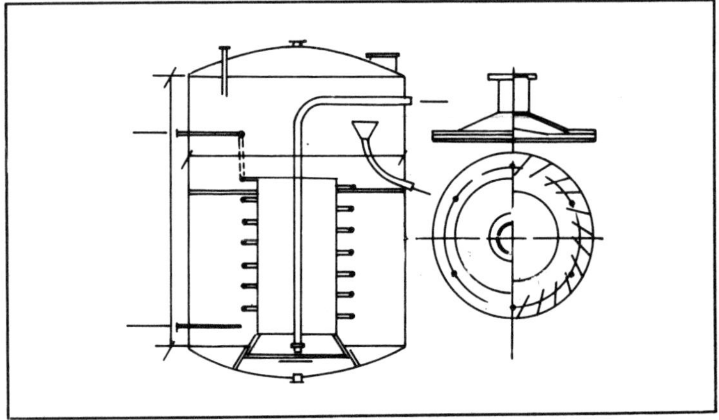

Fig. 11-5. An overflow system.

THE OVERFLOW SYSTEM

The yeast tanks have a pneumatic aerator. Air is pumped in under pressure, rotates a set of vanes, and the overflow spills into a funnel and so on onto the next yeast tank (Fig. 11-5). The overflow system is also similar to that of the fermentation tanks. The primary difference is that in a fermentation tank the idea is to keep things as calm as possible. However, fermentation will cause foaming—especially in a large tank—and to keep the feedstock from leaping out of the fermentation tank a small propellor is installed near the bottom of the tank to gently circulate the feedstock and keep it from foaming (Fig. 11-6).

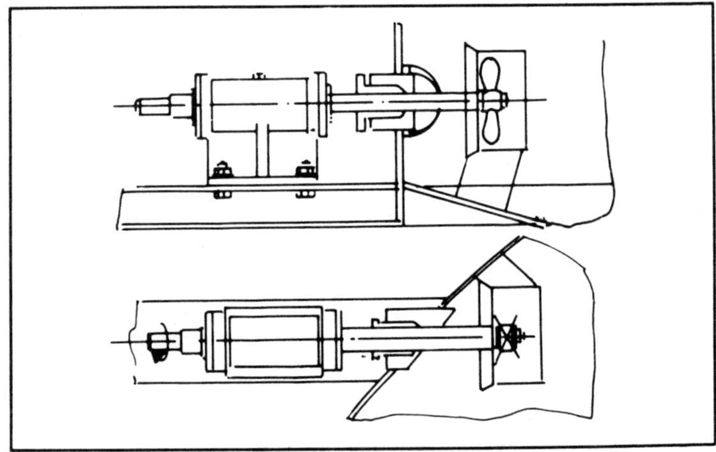

Fig. 11-6. A fermentation tank.

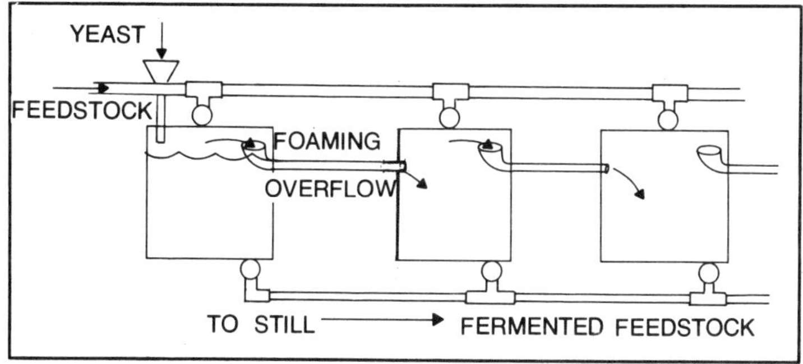

Fig. 11-7. As fermentations reach completion, they are drained out and pumped to a holding tank.

Let's reduce this operation to the 55-gallon drum operation and create a barnyard continuous fermentation operation. But first a few cautionary notes.

If your barnyard feedstock doesn't ferment, there might be insufficient $P_2 0_5$. Before you start—an ounce of prevention is worth a pound of cure—make sure you have some superphosphate on hand. If you have a large series of tanks and don't install propeller mixers, foaming can be suppressed with oleic acid. It's just more expensive.

The concept of continuous fermentation is fairly simple. The pressure of foam is the greatest in the container that has fermented the longest. What happens is that the foam with the yeast overflows into a pipe or series of pipes leading into the next container, starts the process in that container. In turn it becomes more active and overflows into the following container, then the next one after that, and so on. As the fermentations reach completion, they are drained out and pumped to a holding tank prior to distillation.

The overflow pipes are installed in the top of your 55-gallon drums and the drain valves are installed at the bottom. A separate pipe pumps the feedstock into the drums. This is controlled by faucet handles. A typical arrangement would look like Fig. 11-7.

This might seem like an awful lot of bother for a barnyard operation. However, you never know what will come up. One fellow called me from Colorado one night after someone had given him 6000 gallons of apple juice. He hadn't even built a still.

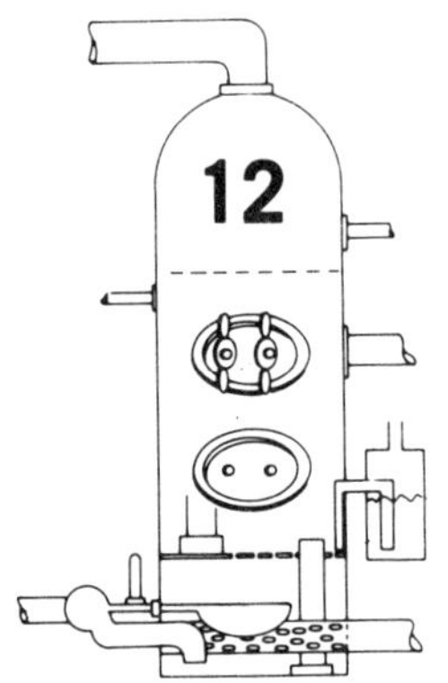

Make Your Own Gasoline

The primary difference in the production of gasoline and the production of alcohol, from starch or other products, is that petroleum occurs in nature while fermentation—though it occurs in nature sporadically—simply does not occur in sufficient quantities without human supervision to warrant any industrial use.

Petroleum and natural gas are found in pockets within the earth's crust from depths of only a few feet to several miles. Normally it is found associated with sedimentary rocks. When a hole is drilled through the rock into such a pocket, petroleum is forced out because of the pressure of the gases within the pocket. This causes the "gusher" often seen with new wells. When the pressure is exhausted, the remaining crude oil is removed by pumping, water injection, or other means.

About 10 percent of the crude oil so extracted consists of natural gasoline, C_5 to C_7 hydrocarbons. The gas in use today is primarily C_5 to C_{11} hydrocarbons mixed together after being run through extensive processing. If you happen to have a small oil well in your backyard, you could theoretically produce your own gasoline. It just wouldn't be very efficient. If you squeezed 10 percent of your crude oil out as gasoline (it wouldn't be very high quality gasoline), that would still give you only 4.2 gallons per barrel. Crude oil "barrels" are 42 gallons. It would be an expensive proposition.

However, economics isn't what this book is about. We'll leave that to the geniuses in the government who got us in the current financial mess. Let's stick to how-to-do-it.

GAS FROM CRUDE

And so for those of you into do-it-yourself gasoline from crude oil, here's how you do it. As in rabbit stew ("first catch a rabbit"), you have to start with crude oil. Crude oil consists of everything from natural gas all the way up to asphalt. All of the fractions of crude oil boil off at different proportions. Natural gas—methane to butane—will boil off in the open atmosphere. Straight-run gasoline boils off between 100° F and 400° F. Kerosene, diesel fuel, and so on boil off at higher temperatures:

—Kerosene: 340 ° F to 620° F.
—Diesel fuel: 530° F to 930° F.
—Lubricating oil: Above 750° F.

Simply heat the crude oil up to the given temperature and those are the products you get. At least up to 750° F, that is what you get. At that temperature, thermal decomposition sets in and nothing seems to work right. You can cure the problem by using vacuum distillation. Simply pump all the air out of your column so that it requires much less heat to do the same job. Pumping the air out reduces atmospheric pressure (normally, the pressure is 14.7 pounds per square inch). A liquid reaches its boiling point when the upward pressure of its molecules equals the downward pressure of the atmosphere. Heat causes—and increases—the upward pressure. If downward pressure is reduced, so then is the amount of heat required to start the proceedings.

In the petroleum industry, more gasoline is obtained from crude oil by *cracking*. To oversimplify it, enough heat and pressure is applied to, for example, a C_{20} molecule to break it up into two C_{10}'s. C_{20} molecules are used for lubricating oils. Why used motor oil isn't processed into gasoline by this process is anyone's guess. It might simply be that no one has gotten around to it.

A petroleum column is similar to the rectifying section of an alcohol column in that the heart of the system depends on the bubble-cap plate (or tray). From the side it looks like Fig. 12-1. The lines represent crude oil. All the arrows going down represent liquid crude oil and the arrows pointing up represent steam and oil vapor. The bubble caps are the two gizmos in the middle. For a more detailed explanation of how they work, see the section on building a column.

A new wrinkle in bubble-cap trays that has been added in the last few years is the *valve tray*. I've never seen one in a whiskey distillery, but they now seem to be quite common in the petroleum industry.

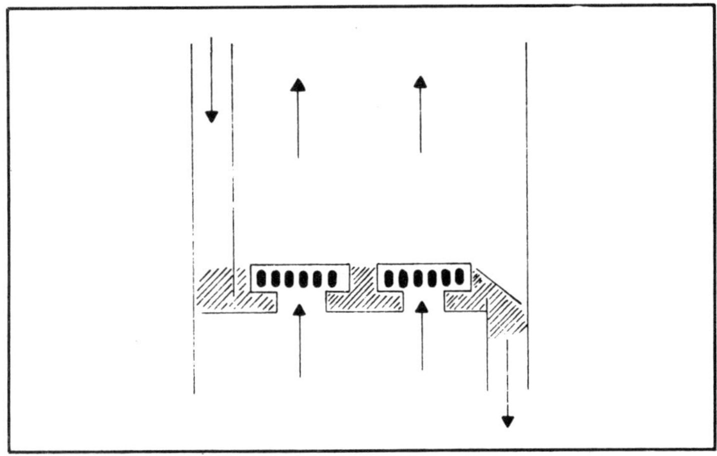

Fig. 12-1. A bubble-cap plate.

Normally, it is a major nuisance to space trays a proper distance apart. The valve tray solves all this by being self-adjusting. It rises with an increase in steam pressure and falls with a decrease.

It works just like it name implies: a valve. A sleeve is first put in the tray (Fig. 12-2). A hollow tube is then fitted to the sleeve (Fig. 12-3).

It should be a smooth fit but not tight. Use the same metal on all these parts or their rate of expansion (different metals) under intense heat is going to foul things up.

Next, fit the bubble cap on the sleeve. Screw threads are desirable because you might want to clean or replace parts on this

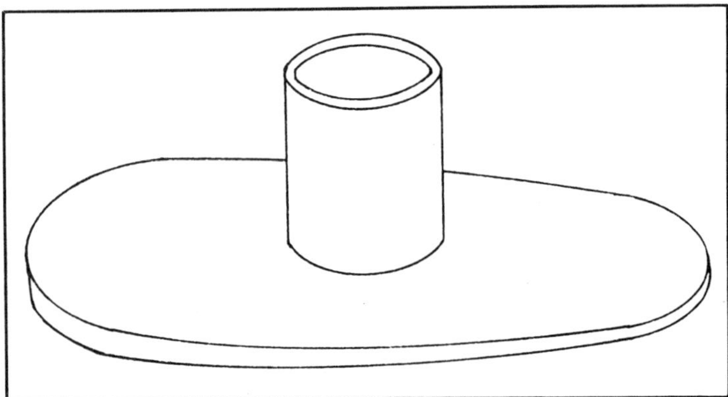

Fig. 12-2. The sleeve.

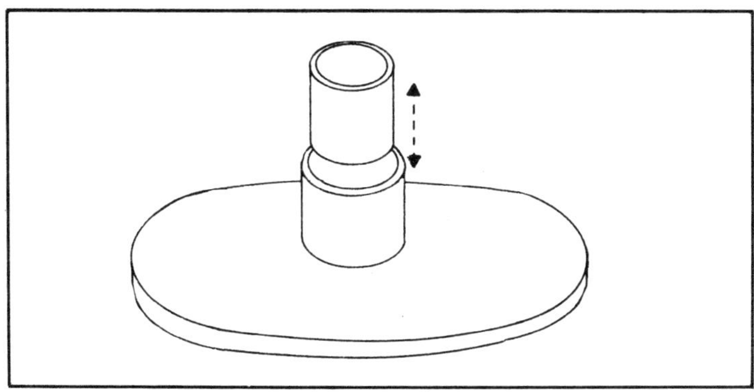

Fig. 12-3. A hollow tube is fitted to the sleeve.

thing. A bubble cap can be made out of a pipe end if you're in a hurry (Fig. 12-4). Just remember that steam has to escape through the drilled or slotted holes. Screw the bubble cap all the way down with

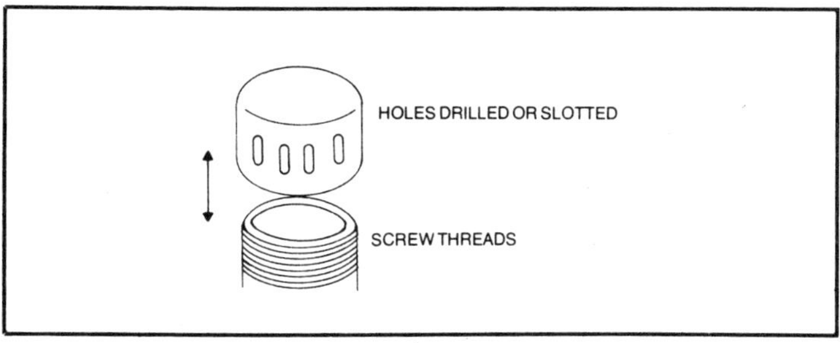

Fig. 12-4. A bubble cap can be made from a pipe end.

no room for the steam to escape and what you have is no longer a still—it's an explosion!

On the bottom end of the sleeve, use either a set of prejections or something similar to a pipe coupling to keep the sleeve from

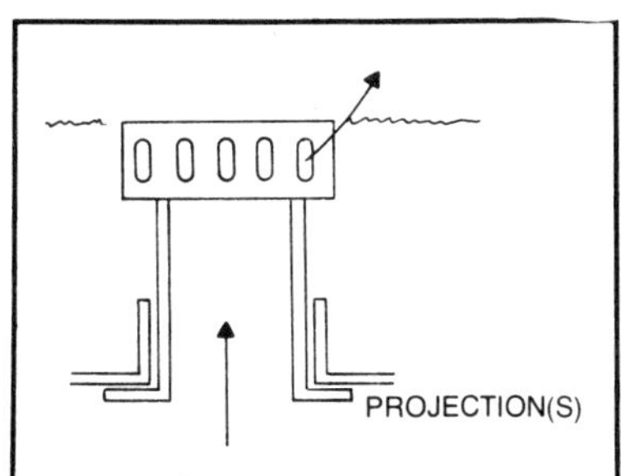

Fig. 12-5. Projections keep the sleeve from rising.

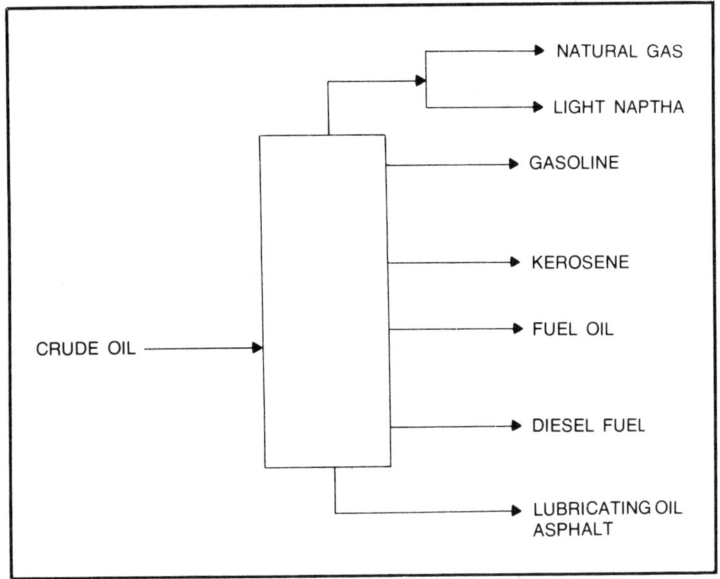

Fig. 12-6. Schematic of a crude fractionating column.

rising all the way up out of the tube (Fig. 12-5). The schematic of your crude fractionating column will look something like Fig. 12-6.

EXPECTATIONS

If you're bound and determined to stick something like this in your backyard, don't expect everything to work perfect the first time you try. It's really a job for a chemical engineer.

If you must, install temperature and pressure gauges at each draw-off level of the column. The crude oil should be pumped in (or gravity fed) about one-third of the way up the column. You can either heat the column with an open flame at the bottom (a risky procedure at best) or simply pump in superheated steam at the crude oil level.

If you do use steam you will have to install steam strippers at the kerosene and fuel oil levels. Otherwise, you will wind up with water in both fractions. This makes it hard to burn.

There will be different heat levels (hotter toward the bottom if you use an open flame) and that explains why different fuels come off at different levels of the column.

13

Alcohol From Cellulose

The conversion of cellulose, such as sawdust, corn stalks, newspaper and other substances, to alcohol is a fairly uncomplicated and straight forward process. At the moment, it is a bit expensive but that is hardly a problem that needs to be addressed here. Just a few years ago the idea of running a car engine on alcohol was preposterous—it was too expensive. Of course, back then gasoline was less than 50 cents a gallon. What might be uneconomical at this writing might be a bargain by the time you read this.

FUEL FROM SAWDUST

Let's say you want to make alcohol from sawdust. There are two types of alcohol you can obtain from wood; methanol and ethanol. Methanol can be obtained from wood by high temperature destructive distillation. Methanol is also known as *wood alcohol*. The other method used to obtain ethanol involves converting the sawdust to simple sugars, the usual fermenting by yeast, and the usual distillation of the fermented solution. There are a couple of other steps involved prior to distillation that are distinct from the standard processes almost everyone is familiar with. To save you the trouble of trying to remember whose book you read last week or where in this one you need to rummage around in for the supporting information, I will provide the usual cookbook instructions.

The first step involves obtaining our standard piece of chemical engineering equipment, the discarded 55-gallon drum. You will need more than one.

The substances you will need to conduct the chemical phase of this operation are sawdust (for example), sulfuric acid, water, and possibly some sodium hydroxide, NaOH.

For the mechanical segment, you will need standard window screens you can buy at the hardware store, plumbing pipes, elbows, couplings, nipples, flanges, and a welding outfit.

I will describe this just the way my partner and I did it in the lab with the exception of some of the plumbing connections. This is necessary because you can't pick up a 55-gallon drum between your thumb and forefinger the way we do a test tube or beaker in the lab.

Be sure that you read all the way to the end of this chapter before you put your hands on the chemicals. You might be unpleasantly surprised.

STEP-BY-STEP PROCEDURES

Pour the sawdust you intend to convert to alcohol into the drum. Don't fill the drum more than one-third full or you will be taking a chance on part of the process slopping over the sides of the drum.

Next, pour what chemists refer to as 18 Molar H_2SO_4, sulfuric acid, over the sawdust. The commercial designation, if you order it from a chemical supply house, would be 100 percent sulfuric acid. However, as low as 91 percent will work. We tried 9.2 Molar, or 51 percent, in the lab and it simply didn't work. It just sat there and looked at us.

Make sure that you put the sawdust in first. If you don't, the sawdust will float on top of the acid—unless you pour in more sawdust than the acid can absorb. In that case, you will simply have to pour in more acid anyway. It's easier to do it right the first time.

When you pour the sulfuric acid on the sawdust, the reaction is almost immediate. The sawdust and acid react in such a fashion as to turn black almost immediately. It resembles an ugly collection of coal tar or pitch. Bubbles rise up through the solution. The bubbling is primarily due to air pockets inside the sawdust. Even though the reaction appears to be instantaneous, you should let the mixture sit for a day or two to allow whatever reaction doesn't take place at once to proceed at its own leisure.

Once the reaction is complete, you can't simply dump in yeast and expect the mixture to ferment. The pH of the mixture is so low, that is the substance is so acidic, that any microorganism such as yeast that you dump in is simply going to explode. Of course, they will be very tiny explosions.

The proper procedure here is to supply enough water to raise the pH to the proper level for fermenting or yeast propagation, 5.0 to 6.0. In Kentucky, where the water is lightly acidic, diluting the solution 50 percent by adding an equal volume of water will raise the pH to about 3.0. In areas where the water tends toward alkalinity (or is basic, in chemical terms) the pH will go higher. If you don't want to keep adding water, add some sodium hydroxide, NaOH, to raise the pH up to optimum conditions. Litmus paper with a matching color chart on the side to indicate pH is available from almost any chemical supply house.

The trick here is that this mixture must be poured into the water used to dilute it with. If you pour the water onto the acid—a natural inclination—what you will get is a loud hissing sound followed by acid vapors rising up out of the solution to attack you. If you add the acid to the water, the dilution factor is much greater. The same reaction will take place but on a much smaller, safer scale.

What takes place is an *exothermic reaction*. That is, large quantities of heat are liberated. You can get a good idea of how much heat is liberated by simply placing your hand on the container during various stages of the proceedings. Briefly. Put your hand on the drum when the sulfuric acid is poured on the sawdust and you will experience the same discomfort that you would if you placed your hand in the middle of a hot frying pan. You will get burned.

Once the solution has been adjusted to the proper pH, it is time to pitch in your yeast. A small packet of Fleischman's, available at the local supermarket, will do just fine. Watch for bubbles of carbon dioxide to appear. They might be hard to recognize coming up through the black gunk; 72 hours, or 3 days, should be enough to allow it to ferment completely.

A word of caution. You might think that simply diluting the acid with half water before you pour it on the sawdust would save a lot of trouble. In a way, it does. You don't have to worry about distillation if you do it like that because 50 percent sulfuric acid won't convert cellulose to sugar and the yeast won't ferment anything else. We tried it in the lab and it simply doesn't work.

LIGNIN

Before you run your solution into your still, you need to get as much of the black gunk, big gobs of it, out of the solution. Remove as much as possible. The material is *lignin* or the substance that bonds sugar molecules together to make cellulose out of them. In a

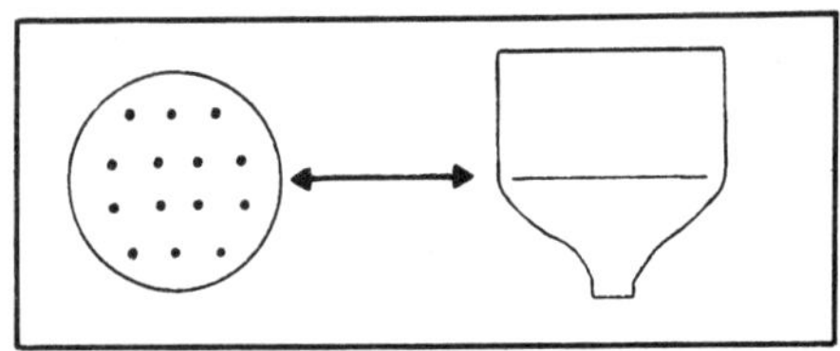

Fig. 13-1. A buchner funnel.

chemistry lab, you use a *buchner funnel* and filter paper. A buchner funnel has tiny holes in the base and looks something like Fig. 13-1.

The filter paper is placed on the bottom, covering the holes, allowing the liquid to pass and trapping practically all the lignin. For a barnyard operation, you can punch nail holes in the bottom of a 55-gallon drum and cover them with newspaper.

Given the fact that the chunks of lignin in an outdoor operation will be much larger than those in a lab, you will probably want to install a series of wire mesh screens between your fermentor and the eventual modified buchner funnel. The screens toward the fermentor should increase in mesh size and those toward the funnel should decrease in mesh size.

The fluid that gets past the newspaper should be yellow in color. The filter won't catch everything. In the lab, we observed a ring of small, brown flakes that settled to the bottom of our distilling flask. This fluid contains ethanol and it is ready to be distilled.

At this point, go back and scrape the lignin off the screens and remove the lignin-saturated paper from your funnel. This is the fuel to fire your still with. There won't be enough to get the whole job done, but it will help and it does eliminate the problem of what to do with all that black gunk. Just be sure you give everything a chance to dry out *before* you try to light it.

The alcohol you get from distilling the yellow fluid is identical to that obtained from sugar or starch. We obtained 190-proof ethanol the first time through a fractionating column. The yield-per-pound appeared to be quite good. According to most of the chemical literature we read prior to conducting this experiment, the commercial yield of cellulose is far inferior to that of corn or other common feedstocks. However, a ton of cellulose (saw dust) is free for the asking.

In place of the sodium hydroxide, NaOH, that we used in the lab, you can substitute common garden variety lye to adjust your pH. If you spill sulfuric acid on yourself—it is a strong acid and it will burn—dilute it with water and scrup with soap. However, the soap should be one that lathers very well because the acid is a very

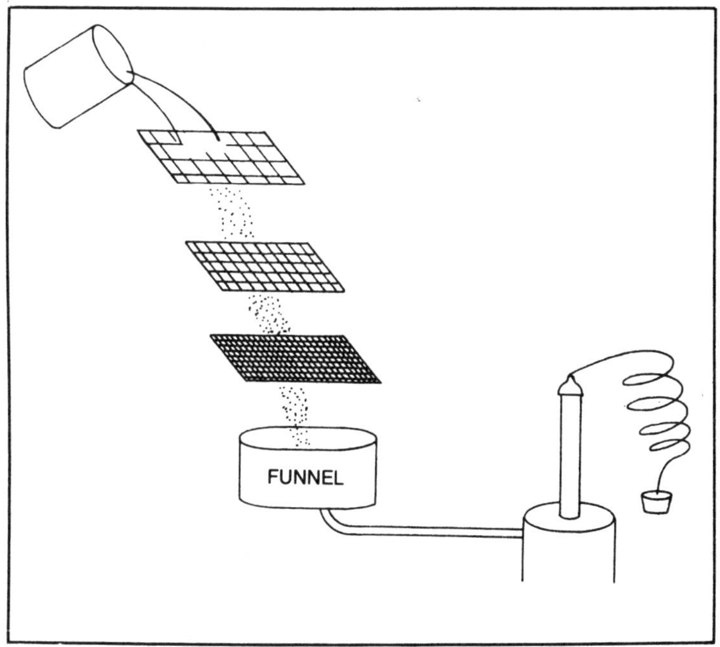

Fig. 13-2. Making alcohol from cellulose.

strong acid and the soap is a very weak base, or neutralizer. Lather the soap up well and use a lot of it.

Once you have distilled the alcohol, you can raise the temperature under your column and boil off the water. Because the sulfuric acid has a much higher boiling point than water, you are simply repeating the distillation process to recover whatever unused sulfuric acid is available from the bottom of your still. You can't recover much of it because H_2SO_4 loses the two hydrogen atoms, or protons, in the initial reaction and is no longer sulfuric acid.

In a commercial plant, the elements involved in the reaction could be recovered in the following fashion. It is a process too long and involved to go into detail here:

$$SO_2 + H_2O \rightarrow H_2SO_4$$

The entire process for this chapter can be diagramed as in Fig. 13-2.

Making Alcohol Fuel Without Distillation

Several years ago, I came across a piece of information I thought a bit unusual: the premium gasoline sold by Arco was supposed to be 7 percent t-butyl alcohol. At the time, I didn't know what either a "t" or the "butyl" stood for.

Butyl is explained in Chapter 19. For your convenience, n-butyl is illustrated in Fig. 14-1. *T-butyl* alcohol means *tertiary butyl*. Figure 14-2 shows the molecular configuration.

Notice that you have the same number of everything; it's just organized differently. If you ever get deeply interested in running organic compounds such as this one in internal combustion engines, you will find that the closer together everything is—such as the t-butyl molecule illustrated—the less likely the compound is to cause detonation in the engine. In fact, that's the way *octane* is rated to gasoline. A straight-chain hexane (6 carbon atoms with hydrogens attached) is given an octane rating of "0" because it hammers and knocks all over the place. Iso-octane—a scrunched-up version of eight carbon atoms with hydrogen attached—is given an octane rating of "100" because it ignites so smoothly.

Some alcohols simply won't work at all as motor fuel; and for reasons no one but a chemist would suspect. The t-butyl alcohol just described won't work, but n-butyl, iso-butyl, sec-butyl, and most of those below a 10-carbon chain alcohol will work just fine. If all the terms confuse you, hang in there and I'll explain.

First, *tertiary* simply means that the carbon bound to the – OH group (which is what makes a hydrocarbon an alcohol) is bound to three other carbons. Look again at the molecular configuation of t-butyl alcohol (Fig. 14-2).

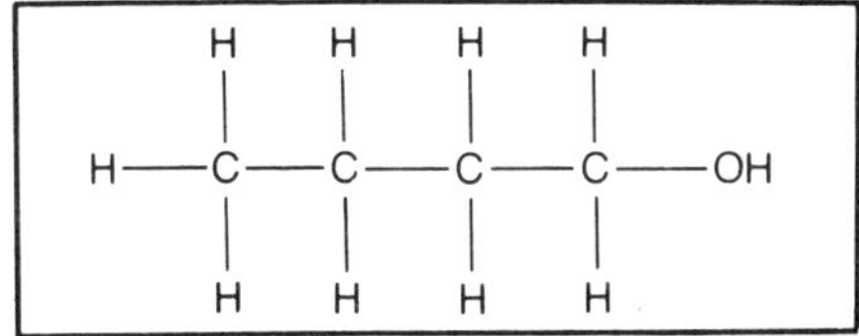

Fig. 14-1. N-butyl.

Sec-butyl simply means the – OH group is on the second carbon in the chain instead of the first. That carbon is in turn bound to two other carbons, making it a "secondary alcohol" (Fig. 14-3).

PRIMARY ALCOHOL

A "primary alcohol" in this series can take one of two forms. It is called "primary" because the – OH group is bound to the end carbon atom and because that carbon atom is in turn bound to only one other carbon. When I use the term "end" that doesn't mean the – OH group is bound to the #4 carbon atom, it simply means it is at one end of the molecule. If the – OH group is at one end of the molecule the IUPAC (International Union of Pure and Applied Chemistry) designation would be 1-butanol—meaning the – OH group is on the no. 1 carbon. Figure 14-3 is for "2-butanol" because the – OH group group is on the no. 2 carbon. In the IUPAC system for naming alcohols, the – OH carbon is always assigned the lowest possible number.

The first primary alcohol is n-butyl (Fig. 14-1). The other primary alcohol is called iso-butyl and looks quite a bit different. However, it is still a primary because the – OH group is still attached to one carbon atom (Fig. 14-4). The carbon, as usual, forms four bonds (just in case you thought I was being decorative).

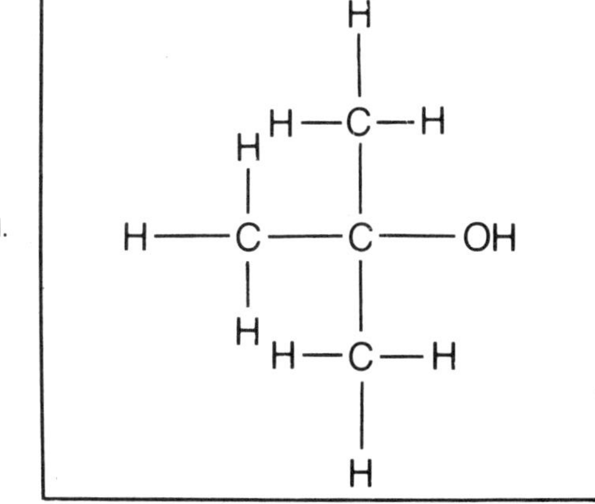

Fig. 14-2. Tertiary butyl.

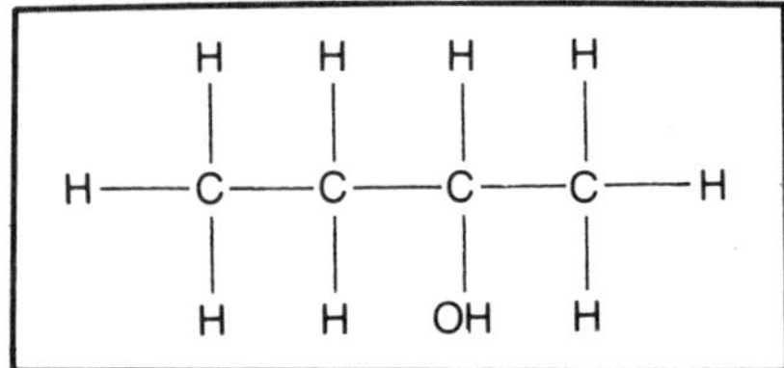

Fig. 14-3. A "secondary alcohol."

What these different configurations cause is different melting and boiling points of what you would *think* are exactly the same things—types of butanols. What the different melting and boiling points cause for you—if you're not aware of them—are problems. Remember, this stuff has to flow through a main jet.

With n-butyl alcohol, there's not a whole lot to worry about because at temperatures under −20° F it's still a liquid. However, t-butyl alcohol—if it's not placed in another solvent to keep it in suspension—will solidify at temperatures over 70° F. You might as well fill your tank up with animal fat if you intend to park in the shade, even on a hot summer day. Once you get above a 10-carbon chain alcohol—such as decanol—the same problem sets back in. The stuff will get solid on you.

MAKING BUTANOL

In this chapter, we'll content ourselves with simply making butanol. And that with only one method. There are a lot of other organic compounds—including but not limited to alcohols—that can be synthesized into motor fuel. The problem we have in including all the other methods is that it would take thousands upon thousands of pages. Organic chemistry is the largest body of organized knowledge in the world. For the moment, let's be content with simply making butanol and getting from point A to point B.

Fig. 14-4. The -OH group is attached to one carbon atom.

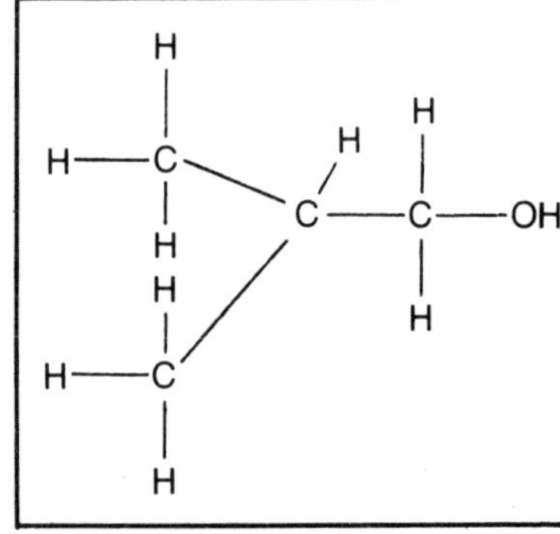

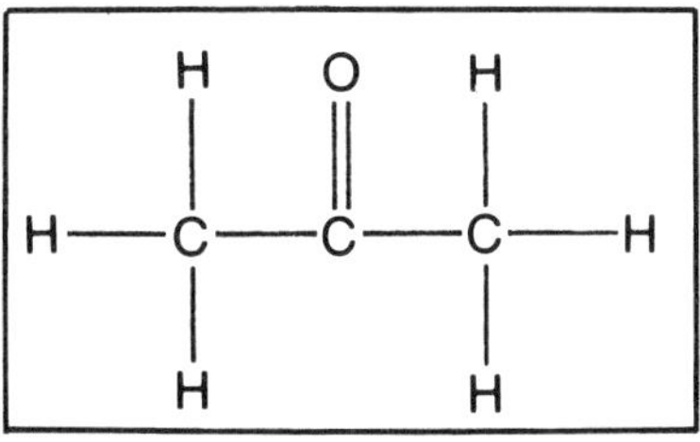

Fig. 14-5. The molecular configuration of acetone.

So how do you make butanol?

It's so simple it's disgusting. In fact, it's so easy that I'm afraid the last of the "how-to-make-alcohol" books has already been written: this one.

The first items on the agenda are three beakers to get the process started. You could start off with a 55-gallon drum, but I wouldn't recommend it unless you have had a lot of practice. Something always goes wrong.

Sterilize the beakers.

Sterilize the drum, too, if you're going to use one.

Allow the beakers to cool.

Dump some water in the beakers. The pH doesn't appear to be important. At least, that was our experience in the lab.

Add your ground-up starch or sugar product to the water. The organism to use is a starch splitter. It completely eliminates the need for yeast or the regular commercial enzymes. I didn't believe this at first—it was much, much too simple—but I did see it work in the lab. This particular organism not only turns the starch into a sugar, it simultaneously converts the sugars into butanol, acetone, and ethanol.

Once you have added the starch to the water, you need to add a bit of ammonia or other source of nitrogen. There is a good chance the ground-up tops of sugar beets or something similar would do.

Next add the organisms. The critters you want are called *Clostridium acetobutylicum* (mouthful, isn't it?) and can be ordered from:

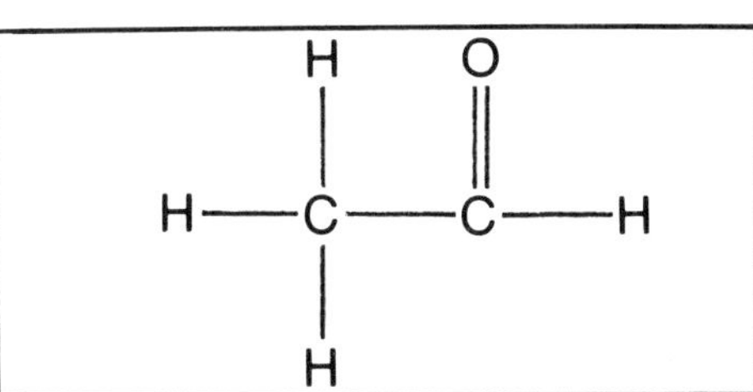

Fig. 14-6. An aldehyde.

Presque Isle Cultures
P.O. Box 8191
Presque Isle, Pennsylvania 16505

This outfit is sort of a "mom and pop" operation and the folks who run it are extremely conscientious. They won't ship an order unless they know the microorganisms are thriving. What you get from them is the organisms growing in a test tube full of nutrient jelly.

Stick a spatula down in the test tube and pull out a chunk of jelly. Plop the jelly in your beaker. Temperature control is important. At 98.6° F, everything works according to Hoyle, so to speak. At this temperature, the Clostridium acetobutylicum does everything it is supposed to. Butanol floats to the top while acetone and ethanol remain in the water.

If you see little black spots in your solution within a few hours, then you are on your way. Within a 24-hour period, you should see a layer of butanol float to the top of the beaker. At the bottom of the butanol layer, you will be able to observe a collection of very tiny, sharp bubbles—hydrogen and carbon dioxide gas.

The reason for using three beakers is that sometimes the organisms don't take hold or they die off. In our lab experiments, we only got two out of three to work. If you discover any problem with sanitation, you can forget it.

Once you have the beaker working properly, dump it into a 55-gallon drum full of starch or sugar and see if you can get it to continue the fermentation process. If you can, you have arrived. If you can't, throw in some more organisms. If that doesn't work, clean up and start over. You may have contaminated your mixture with another microorganism.

TEMPERATURE CONTROL

Temperature control is important to maximize the yield in a minimum amount of time. At 53° F the critters won't work at all; that's the temperature you store the stuff at. How you kill them off is as follows (it's difficult).

The only way that spore-forming species such as those of the genus Clostridium can be killed using heat is through the process of *tyndallization*. Tyndallization is a process of fractional sterilization. The culture of spore-forming microorganism is exposed to heat (boiling water) for 30 minutes each day for three *successive* days to kill all the cells.

The reason that heating for only one day fails is that the heat drives the vegetative cells into the spore stage. It is true that some cells are killed before they enter the spore stage, but some enter it nevertheless. So after heating one time, there are still enough cells left to grow after the conditions have returned to optimum. Tyndallization takes this into account. The medium returns to normal and the cells become vegetative, but this takes 18-24 hours, and at that time the heat is reapplied. This kills additional organisms. This cycle is kept up until no more cells are living. Remember that no matter how hot (short of burning) you get the microbes with one heating, they are are not all destroyed.

The vegetative phase leads to an active growth period. The primary damage done to such an operation is by other, uncontrolled, outside microorganisms.

Back to the cookbook instructions. If you see the butanol layer rise to the top, just skim it off, recalibrate your main jet or jets, and dump it in your fuel tank. The butanol will get up to 160-proof on the top layer if you pour some salt in the solution. The salt will dissociate in water into sodium and chloride, but not into butanol. Water is a highly polar (molecules attract each other) medium and butanol is an almost totally non-polar solvent.

ACETONE

You might want to recover the acetone produced. What you get out of the feedstock is 6 parts butanol, 3 parts ethanol, and 1 part ethanol. Acetone is extremely useful for paint thinner, a base for fingernail polish, and is a key ingredient in the manufacture of smokeless gunpowder.

The molecular configuration of acetone looks like Fig. 14-5. The double bond to the oxygen atom with methyl groups on each end is common to a whole family of chemicals known as *ketones*. Next time you hear or read of someone from the auto or oil companies pontificating about "aldehydes and ketones" in the exhaust emissions of alcohol-powered vehicles, you'll know what they're talking about. An aldehyde is identical to a ketone with the exception of a hydrogen that replaces the methyl group on one end of the molecule (Fig. 14-6).

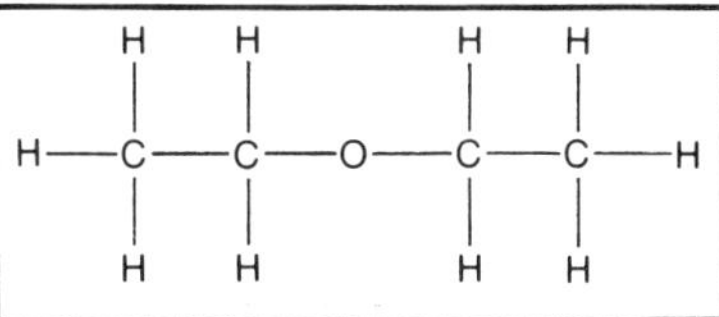

Fig. 14-7. The molecular configuration of ether.

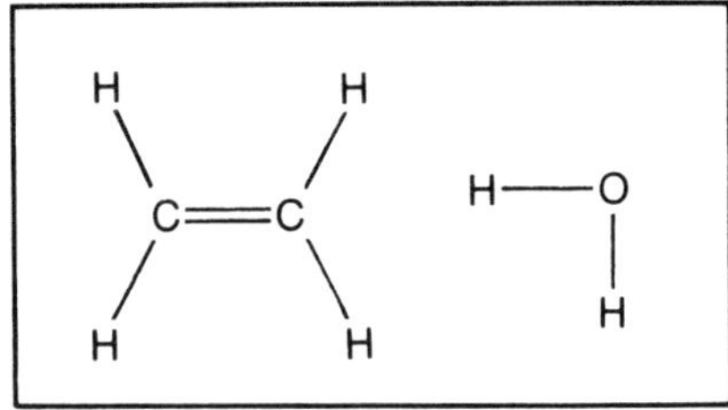

Fig. 14-8. Ethene gas and water.

ETHER

If you stick the oxygen in the middle of the molecule, what you wind up with is an ether (Fig. 14-7). In case you're wondering what value ether has, it's used as a "starting fluid" for gasoline and diesel engines in extremely cold weather. You can make your own starting fluid for your alcohol-powered vehicle simply by mixing sulfuric acid—H_2SO_4—with ethanol at 284° F. You will get some water with it but it will work. However, if you cook this solution off at 338° F you will get ethane gas and water (Fig. 14-8). Ethane is a vapor at room temperature. At 32° F you will get ethyl hydrogen sulfate.

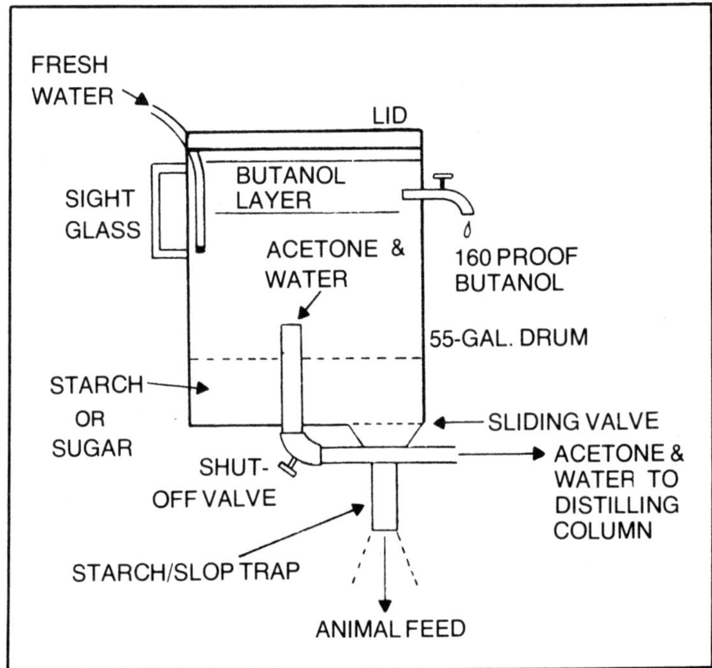

Fig. 14-9. A sample motor-fuel/chemical plant.

The easiest way to run this operation is to just throw the starch and microorganisms in 98.6° F water and then skim the butanol off the top as you need it. The beautiful part of this process is the low temperature required. You can stick this in a drum in your basement with an immersion heater in it or paint the drum black and put it out in the sunlight. The acetone boils out of the solution at slightly over 133 F. This makes it fairly easy to collect. A few words of caution are in order:

☐ Acetone evaporates readily. Keep the collected acetone covered.

☐ Hydrogen gas is a byproduct of butanol production using this method. Use an outside vent to avoid a fire.

☐ The ether you make from ethanol and sulfuric acid is the same stuff used as an anaesthetic. Don't put yourself to sleep.

A sample motor-fuel/chemical plant is shown in Fig. 14-9. It can be operated on a continuous basis. If you want to extract the butanol remaining in the water, it can be boiled off at 243° F. You can boil or distill this mess in three separate operations and butanol is left over:

Acetone: 133°F.
Ethanol: 173° F.
Water: 212° F.

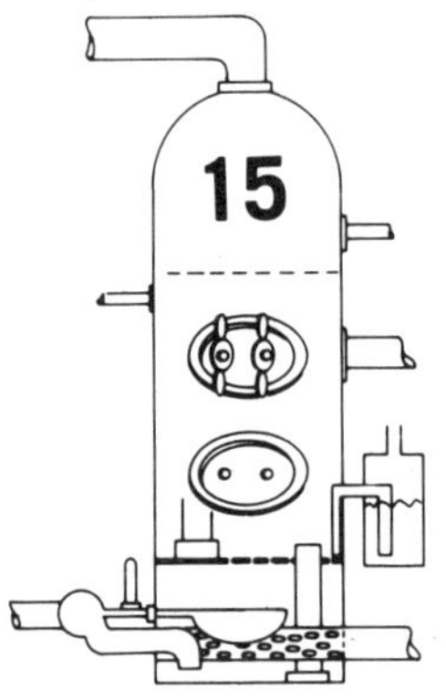

Extracting Moisture From Grain

From every bushel of corn you can get approximately 18 pounds each of alcohol, carbon dioxide, and cattle feed (known in the trade as *distillers dried grains*).

I say "can get" because if you don't use the slop for cattle feed within three days, it will go sour on you. It is possible to heat the tanks the slop is held in and keep the stuff usable. After that, wet stillage simply isn't any good for anying.

If you remove the moisture—dry it out—the dried grains will keep almost indefinitely. In addition to use as feed for cattle and chickens, dried grains can be burned to provide potash, scrubbed to provide ammonia, and for a number of other products.

AMMONIA EXTRACTION

Start with a simple scrubbing operation to extract the ammonia and see how it works. Don't even try to purify the ammonia-water mixture once you scrub the wet slop because ammonia boils at temperatures well below 0°F and there simply isn't any way you're going to capture it and keep it in a barnyard operation. There really isn't any need to because the water-containing ammonia can simply be pumped out to your garden or farm lands to replace the nitrogen in the soil and take care of irrigation at the same time.

Scrubbing can also be used to separate the final dab of water from alcohol so that the alcohol can be blended with gasoline. More than 4 percent water in a gallon of alcohol added to 9 gallons of gasoline (as almost everyone now knows, this is called *gasohol*)

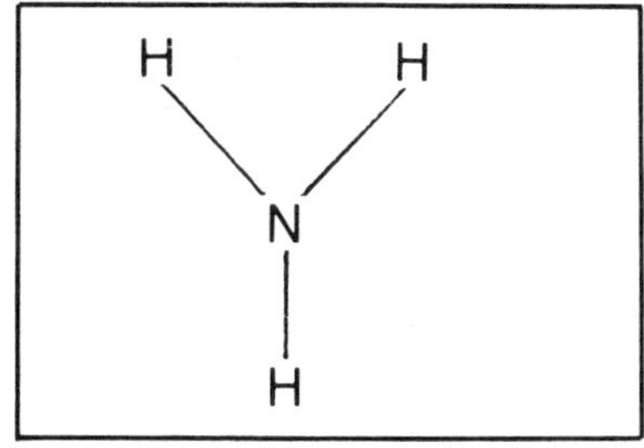

Fig. 15-1. The structure of an ammonia molecule.

will cause the gasoline and the alcohol to separate. This makes for a host of problems in starting and running your engine.

Ammonia in very small amounts will not escape the water it is trapped in. In large amounts, it leaves in a hurry; that's the smell that bowls you over when you open a bottle of cleaning fluid. What keeps the ammonia trapped in water is hydrogen bonding. That subject is covered in the Chapter 19. Figure 15-1 shows the structure of an ammonia molecule. The "N" stands for nitrogen.

To extract the ammonia from the wet slop, you simply have to boil water and then introduce the resulting steam into used condenser water and send it on its way. A sample setup will look something like Fig. 15-2.

The ammonia and water (or steam) boils off and enters the small tank. The used liquid water leaves the top of the cooling jacket around the column (which is not packed for this operation) and enters the small drum or tank simultaneously with the ammonia/steam mixture. The cooling water traps the ammonia and keeps it in suspension until it reaches the soil. Or wherever it's going.

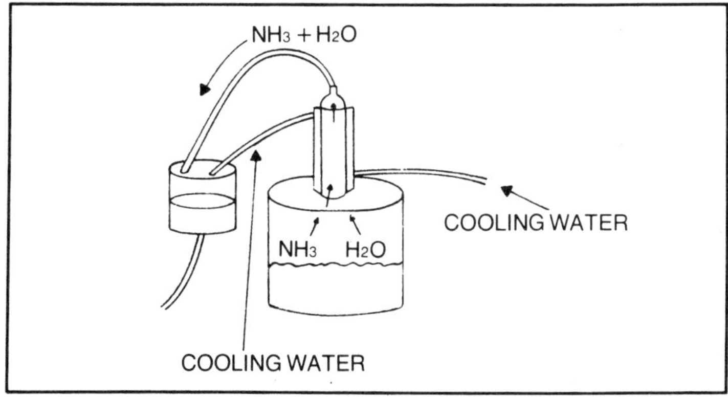

Fig. 15-2. Extracting ammonia.

If a group of farmers got together, they could probably set up an ammonia plant. However, it's unlikely they could do it with old 55-gallon drums. Ammonia is often used for air-conditioning and cooling equipment. This type of equipment is carefully machined and fitted and often it still leaks.

In Fig. 15-2, the direct source of heat is not needed if you use effect stations and simply extract the vapor from them as described later in this chapter.

GASOHOL

A principle similar to this can be used to obtain almost 200-proof alcohol that is suitable for dumping directly into your own gasoline tank. There have been a lot of claims by the makers of table-top stills and the like touting "make your own gasohol," but if you read the fine print they're talking 170-proof or so. Totally unsuitable!

It's my opinion that the whole gasohol business is only an oil company engineered program to keep the farmers out of direct competition with big business. But for those of you who still think 1 gallon out of 10 will do any good, here's how you do it.

Start with the usual 55-gallon drum. The wrinkle here is that you need a 90-degree elbow and a condenser without any coils in it. See Fig. 15-3. What happens is that the alcohol/water mixture, containing some benzene, is boiled off and makes a 90° turn. The column to the right of the drum contains a layer of benzene that is level with the bottom of the horizontal pipe.

At 68°F, benzene will absorb only 0.06 percent water. What happens is that the water and benzene condense in the condenser and fall back down into the vertical pipe. The benzene, being lighter than water, floats to the top. The water sinks to the bottom of the trap. If you want to be technical, this is known as a *Dean-Stark trap*.

As the water and benzene fall back into the trap, the water sinks but the benzene that rises past the level of the horizontal pipe overflows into the boiler and the entire process repeats itself. Periodically the water should be drained from the bottom of the trap. Theoretically, benzene can cause cancer but so can tobacco—only quicker and surer. So who cares?

So much for scrubbing operations. Now, back to the main subject. When you want to procure a liquid from a substance, the process used is called distillation. When you simply want to boil off a liquid and collect the leftovers, the process is called *evaporation*.

Another procedure for getting rid of liquid is called *sublimation*. I will cover it first.

The process by which a solid changes directly to a vapor without turning to a liquid first is called sublimation. Ice to water to steam is a process we are all familiar with. If the ice turns directly to steam—or water vapor—that is sublimation. You might have noticed this happen on a cold, dry day. If you want to remove water by this process, you simply lower the temperature below freezing and create a vacuum for the ice crystals formed to sublimate in. In industry, this is known as *freeze-drying*.

In the distillation industry, evaporation is the more common method. However, it ought to be fairly simple to rig up a freezer and a vacuum pump to create your own cattle-feed drying operation.

EVAPORATION METHOD

To run an evaporator room or operation, you will have to start with two lines—one for steam and one for the slop. A long steam line will "grow" on you in a lengthwise fashion. This is compensated for by a device resembling an accordian placed in the line. In an outdoor operation, the slop line is buried below the frost line to keep it from freezing.

Let's take it by the numbers once the slop reaches your evaporator room. These are only the basics and the principles. Use a little imagination and junkyard your own creation together.

The first step is to centrifuge the slop. The heaviest particles settle to the outside of the centrifuge. The remaining liquid is pumped to feed tanks. The coarse particles are routed to a dry feed container. There they will eventually be mixed in with the syrup you are about to create. Once it is mixed together, the partially dried slop is dried completely in a heated vacuum chamber.

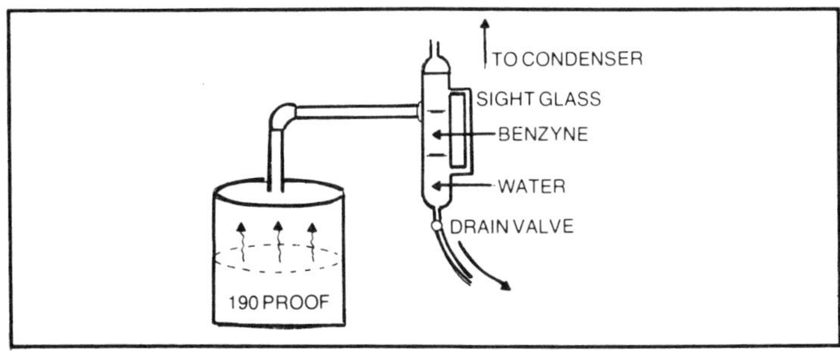

Fig. 15-3. Making 170-proof alcohol.

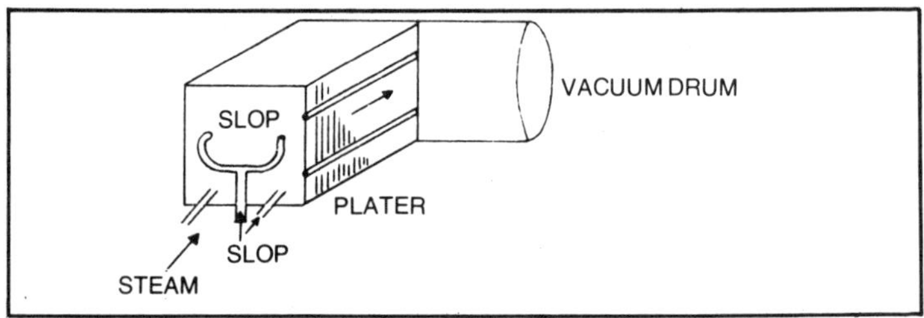

Fig. 15-4. An effect station.

Most of the drying of the uncentrifugal portion takes place in what are known as effect stations: first, second and third effect. The first effect is fed the liquid slurry; part of it is recirculated and part of it goes to the third effect.

The second effect recirculates part of the liquid slurry and routes the rest to what is known as a 25 percent tank (25 percent grains, 75 percent water). The third effect simply removes what moisture it can and recirculates everything else.

Each effect, or effect station, consists of a feed for slop, a feed for steam, a vacuum drum, and a series of feed plates. The effect stations look something like Fig. 15-4.

FEED PLATES

The key to the way an effect station works is in the plates (called *feed plates*). The feed plates each have a small opening on each side for the slurry to feed through. In Fig. 15-5, the two small openings are the ones with the arrows pointing to them at the bottom section of feed plate.

Except for the two small holes, each hole in the plate is completely surrounded by a gasket. Each has the gasket cut away on the inside to allow the slop to flow up the plate (Fig. 15-6). The gaskets beneath the two slop holes keep liquid from falling. Another plate—they are normally packed four together—also helps keep the liquid from falling.

Fig. 15-5. Feed plates.

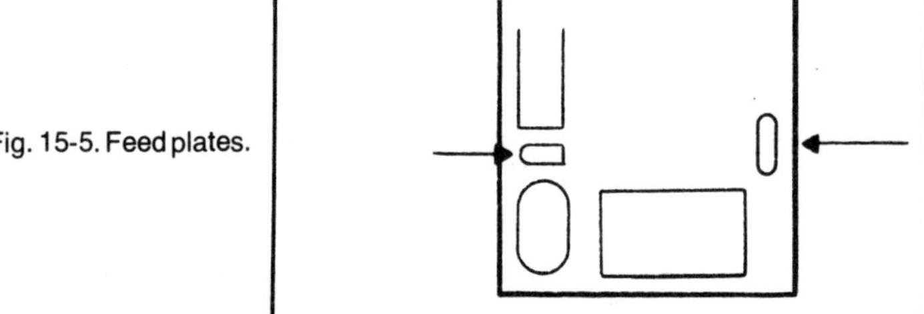

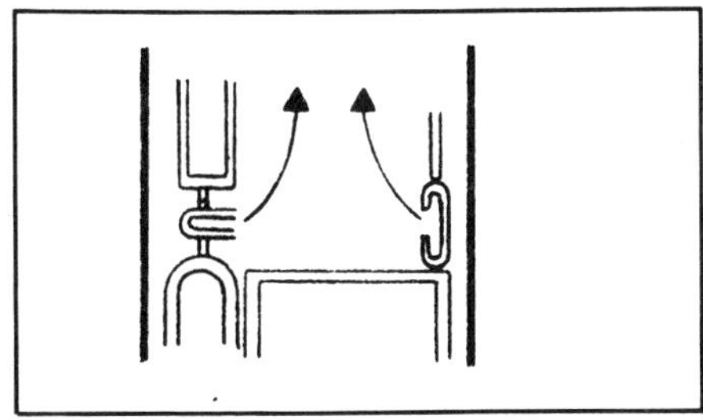

Fig. 15-6. Cut gaskets.

Steam is run between some of the plates to evaporate the water—some of it, anyway—in the slop. After every fourth plate, the liquid falls down the plates into the two large holes on the side and is fed to the vacuum drum. From there it goes to the next effect for recirculation or extraction. It is conveyed by means of pumps (see Fig. 15-7).

These plates have little bumps or studs on them to keep them from being drawn or stuck together because of the vacuum. They do get nasty fairly quickly and they have to be taken out every week or two for a scrubbing down.

Steam is run only into the first effect; the other two pick up enough heat from the product. The slurry from the 25 percent tank, where the effect stations dump their handiwork, is then mixed in with the coarse particles and rotated with an auger in order for the dry grain to absorb some of the moisture from the syrup.

From there, the syrup and dry grain are sucked up into a heated vacuum tower, completely dried, and conveyed to storage towers. This stuff is chock full of vitamins and minerals and almost totally free of carbohydrate. It might even make a great health food if it weren't for the taste. It resembles very heavily salted sawdust.

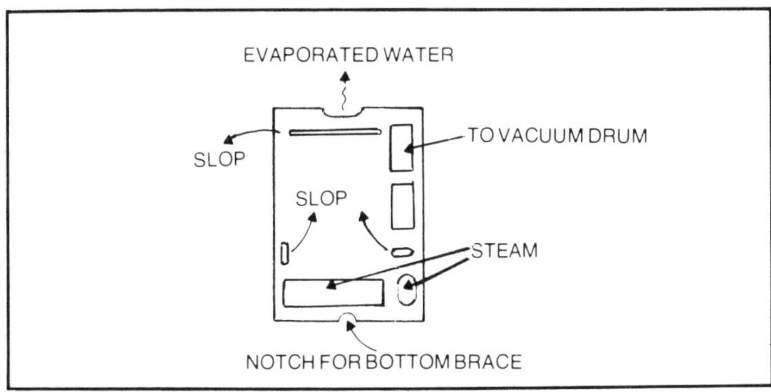

Fig. 15-7. Pumps are used for recirculation.

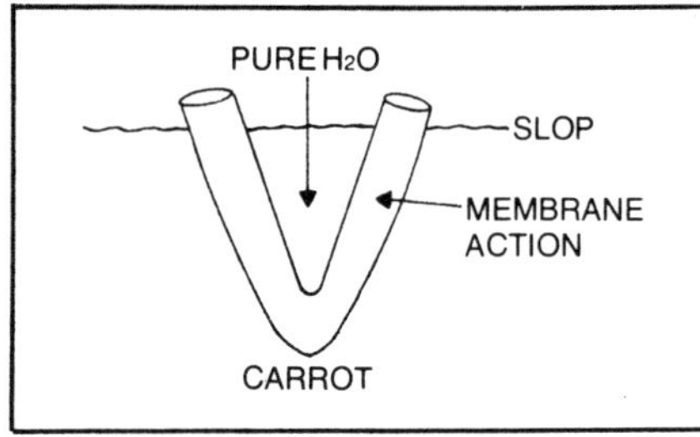

Fig. 15-8. Reverse osmosis.

Once the leftovers have been dried, they can be burned to provide potash, superphosphate, and the oxides of silicon, iron, calcium, magnesium, and potassium. It makes great fertilizer. The reason you get oxides (i.e., the molecules contain oxygen) is that most of the metals mentioned simply do not occur in nature. They have to be purified in a laboratory.

A process being worked on at the moment is membrane separation of the water from the byproduct. So far, nobody has had much luck. The problem is that most membranes allow pure water into a solution of byproduct, a product known as *osmosis*.

Reverse osmosis is the process of separating water from the solution. It seems to work well in nature. For an example, you can stick a hollowed-out carrot in your slop and pure water will flow to the inside of the carrot (Fig. 15-8). The primary problem here is that no one seems to know exactly how the membrance action of a carrot works. It might get a little tedious sticking carrots into your slop.

Dry Ice From CO_2

Most of the "hands-on" enthusiasts getting into alcohol production these days are making no allowances whatsoever for the utilization of all the CO_2 generated in the production of alcohol. Admittedly, there is not the market for dry ice—solid carbon dioxide—that there was 20 or 30 years ago. It is my contention that it is better to plan ahead. While there might not be a large market for dry ice at the moment, sooner or later some young genius is going to think up a use for it that triples its value. A similar situation developed in the petroleum industry in the late 19th century with a byproduct of kerosene production: gasoline.

The production of dry ice involves converting the gaseous CO_2 to a liquid and subsequently to a solid. The key to the system is a three-stage compressor. The gas is compressed in successive stages until it reaches the desired consistency. It can't be done all at once because CO_2 heats up when it is compressed and has to be cooled between compressings.

To give you an idea of how the process works, I'll describe several ways of doing things. This chapter won't be quite as thorough as some of the other chapters; this subject actually requires a book in itself.

TURNING CARBON DIOXIDE INTO LIQUID

Several stages are used in this process. The CO_2 is compressed many times. On the second stage it is cooled. Then it is purified and dried. The carbon dioxide, which is obtained according to one method, is sucked in by means of the first stage of a three

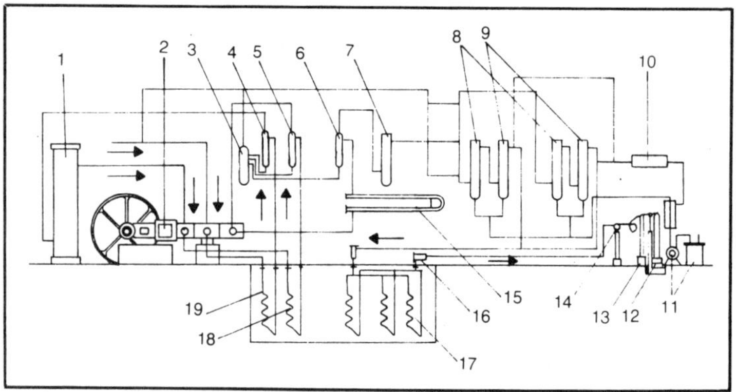

Fig. 16-1. Turning carbon dioxide into liquid.

stage compressor (Fig. 16-1) and is condensed to the density of 3.5 to 4 atmospheres. From there it is pushed into the intermediary cooler 19 and from there into chamber 4 where the oils are separated from the carbon dioxide. From chamber 4, the cool carbon dioxide gas is sucked into the calcium chloride chamber 1 by the second stage of the compressor. It is then compressed to 12 to 18 atmospheres and pushed into another intermediary cooler (18) as well as into the oil separator (5). Next, the gas is sucked by means of the third stage of the compressor, compressed to 60 to 70 atmospheres, and then passes into the cooler (15), oil separater (6), the high pressure filter (7), purifier column (8), and drier column (9). From the drier column the gas is directed through the filter (16) to the condenser (17).

The fluid carbon dioxide from the condenser is then directed to the collector station (14) in order to pour into the balloons (12), in which it is stored and then transported. The scales (13) are used to determine how much each of the balloons weighs.

The oils from the oil separators (4 through 6) are gathered into oil collectors (3). The air that is necessary for the drying purifiers (8) and that of drying column (9) is warmed up in an electrically equipped drying and warming up chamber (10) and then is sucked in by the vacuum pump (11).

This particular method is a common one. Depending on where you obtain the carbon dioxide you can add additional apparatus for its purification and other stages.

LIQUIDIFICATION OF CARBON DIOXIDE

Experiments show that the gases which are produced during fermentation in airtight fermentation chambers are almost pure carbon dioxide containing 99 percent to 99.5 percent of carbonic acid, or CO_2. Another product in this carbon dioxide you find is .4 percent to .8 percent of alcohol according to weight of CO_2, either from 0.3 percent to 0.4 percent, acids .08 percent to .09 percent and some traces of aldehydes. For the method of the liquidification of the carbon dioxide which is produced during fermentation, you must take into consideration the possibility of the absorbtion of the above mentioned elements.

According to the chemical formula of alcohol fermentation, the amount of carbon dioxide is equal to 95.5 percent from the weight of the alcohol. Consequently, for an alcohol producing enterprise whose capability is 8000 gallons for a 24 hour period, the total amount of the carbon dioxide produced is 50,000 pounds. From this number, about 30 percent, or 15,000 pounds, of carbon dioxide is released in the yeast generators and cannot be utilized because it is mixed with the air which is brought in for the purpose of the multiplication of yeast. The degree of utilization of carbon dioxide which is released from the fermentation tanks reaches 90 percent. Consequently, at an alcohol producing plant of the indicated capacity it is practically possible to turn carbon dioxide into liquid in the amount of 13 to 13½ tons. Some alcohol plants have the instruments, shown in Fig. 16-2, for that purpose.

The carbon dioxide released from the airtight fermentation chamber (1) enters into the alcohol catcher (2) where it is rinsed, freed of alcohol vapors, and then enters the gas collector or gas holder (3). From 3, the gas is sucked in by the first stage of the compressor (7) and compressed to 8 to 10 atmospheres. The compressor is operated by the means of an electric motor (6). When you compress the gas it becomes hot therefore, after the first stage of compression it is cooled off in the cooler (11) and then

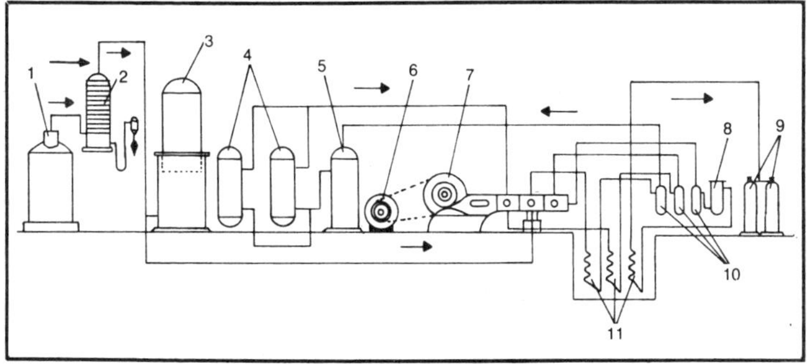

Fig. 16-2. Liquefaction of carbon dioxide.

freed from oil particles in the oil separater (10). The gas is then cleaned of specific smelling substances and other oxidation products which are characteristic of the acids produced during fermentation.

Filtration is conducted by means of a solution of hydrochloric acid and calcium in the filter (5) and is activated by means of coal in one of the chambers (4) which are placed parallel to each other. The filtered gas is now subjected to gradual condensation to 20 atmospheres in the second stage after 70 atmospheres in the third stage. After each of the stages, the condensed gas passes through the cooler end the oil separator. After the third stage, it passes through additional filter 8 and passes through silica gel where it literally frees itself from the remaining humidity. The gas condensed to 70 atmospheres is cooled in the condenser of the third stage to 59°F and turns into liquid. The liquid obtained is poured into containers made out of steel.

PRODUCTION OF SOLID CARBON DIOXIDE (DRY ICE)

The liquid carbon dioxide can be turned into a solid by means of two methods: by means of freezing with the help of a cooler or by means of reduction of pressure. To lower its pressure by using a vacuum is also known as *throttle control.* In the first case, the carbon dioxide turns into a transparent glass-like solid body with the density of about 1½ grams per cubic inch. In the second case, evaporation takes place for up to 80 percent of the total carbon dioxide. The temperature of the remaining part drops and the carbon dioxide in the process turns into a solid body which looks like a mass of snow.

The production of solid carbon dioxide in industry is based on the second principle. As a rule, a frosted white color and its appearance resembling chalk has a specific gravity, depending on the method of production, that fluctuates from 1.3 to 1.6. When heat is introduced, the solid carbon dioxide evaporates or sublimates and turns directly into gas, bypassing the liquid stage. This is where the term *dry ice* comes from.

The solid carbon dioxide has the quality of low temperature production. It produces quite a bit of low temperature or cold. It cools three times lower than ice mixed with salt. In addition, it has positive applications such as in the use of medicines when it is necessary to keep things cool. Absence of moisture helps in keeping things cold and dry.

In temperatures and pressures above five atmospheres, dry ice melts and a liquid carbon dioxide is obtained. In addition to that, it differs from the ice obtained from water in that the volume of the obtained carbon dioxide increases 28.5 percent. You should take that into consideration when you are building and exploiting the necessary apparatus.

The production of dry ice can be accomplished by means of little and low pressure and this depends on the pressure which is used to throttle the liquid carbon dioxide. Figure 16-3 illustrates the production of carbon dioxide by the means of high pressure.

The liquid carbon dioxide obtained in the condenser (15) of the third stage of the compressor (1) and then filtered in the reservoirs (2) under the pressure of 60 to 70 atmospheres is directed to the receivers for the liquid carbon dioxide (3). Then this mass passes through the internal pipes of the first and second section of the heat exchanger (14) and by means of the regulating throttle process to the 24 to 28 atmospheric pressures part of the carbon dioxide evaporates; the temperature of the remaining part lowers to −12 to −8 degrees. The liquid carbon dioxide is built up in this process in the first transitory container (4) and the carbon dioxide vapors produced in the process are separated in this container and then sucked out through a cyclic or round opening. The first section of the heat exchanger (14) is controlled by means of the high pressure cylinder of the additional compressor (13). The level of the carbon dioxide in the first transitory chamber is controlled by means of the mercury indicator (9).

The pressure of the carbon dioxide fluid is lowered from 28 to 8 atmospheres by means of a regulating valve and this again brings about the evaporation of a part of the fluid that causes part of the fluid to evaporate; the mixture of fluid carbon dioxide and gas

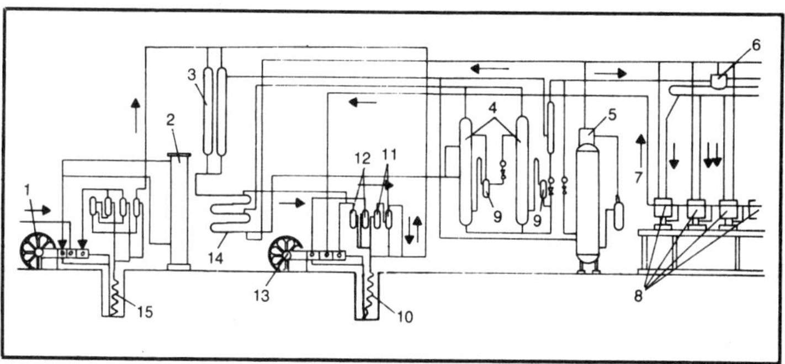

Fig. 16-3. Using high pressure to produce carbon dioxide.

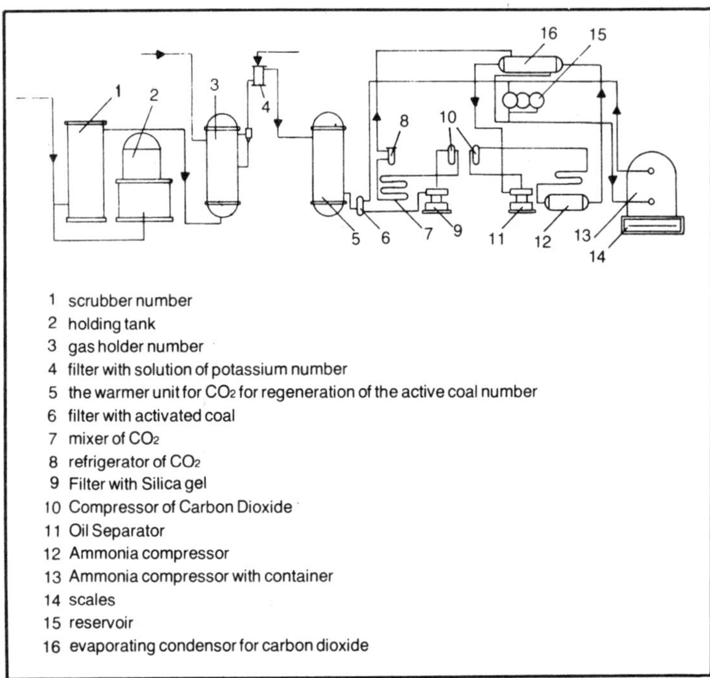

Fig. 16-4. Producing carbon dioxide.

vapor, the temperature of which minus 50°F is introduced in the second transitory chamber (6). Here the liquid separates itself and the vapor is sucked out through the second section of the heat exchanger (14). It is sucked out by means of additional compressor (13), using middle pressure. The level of the fluid in container (6) is controlled by means of light indicator (7). From the second transitory chamber, the fluid carbon dioxide under the pressure of 8 atmospheres, with the temperature – 50°F, is introduced into the ice generators (8), which are turned on in sequence.

When the ice generator is filled, the lower diaphragm opens very slowly. The liquid carbon dioxide, while it passes through the diaphragm, loses pressure. And having again reached the triple point (5 atmospheres) of atmospheric pressure it slowly turns into a solid body. The crystals of this substance fill the whole diaphragm. It is the process by which the block of carbon dioxide built up takes place in the chamber.

First, the diaphragm is covered with the solid carbon dioxide dry ice and from there the dry ice is added to until the entire chamber is filled with one block of ice. It spreads from the bottom

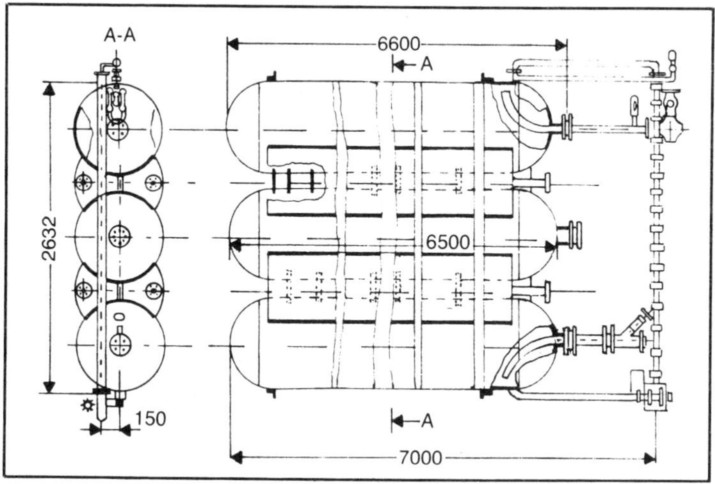

Fig. 16-5. A three-chamber reservoir for storage of liquid carbon dioxide without insulation.

up in a concentric manner. At the end of the process you get a block of dry ice. The weight of one such block is from 92 to 97 pounds. The remainder of the carbon dioxide is stored under the pressure of eight atmospheres in the container (5) and is controlled by means of light indicator level (7).

The process of ice formation is accompanied with production of vapors under almost atmospheric pressure and a temperature of 175°F. These vapors, having passed the diaphragm, are introduced in the jacket of the ice generator and from there by means of pipes are directed into the cylinder of low pressure of the conditional compresser (13). Then they pass through the battery filters and oil separators (12), are compressed to 65 to 70 atmospheres, are turned into liquid in condenser (10), and then they are directed to the battery of receivers (3) for the purpose of repeating the cycle.

The production of solid carbon dioxide at alcohol factories is economically justified because the method, apparatus and technological methods of the production of the dry ice do not depend on the type of the raw material from which the fluid carbon dioxide has been obtained. Therefore, the cost of the solid carbon dioxide is determined mainly by the cost of liquidification of the carbon dioxide gas formed.

Another factor which determines the cost is the losses occured in producing dry ice. In order to produce 1 pound of solid carbon dioxide, you have to use up from 1.6 to 2 pounds of liquid carbon dioxide. See Figs. 16-4 and 16-5.

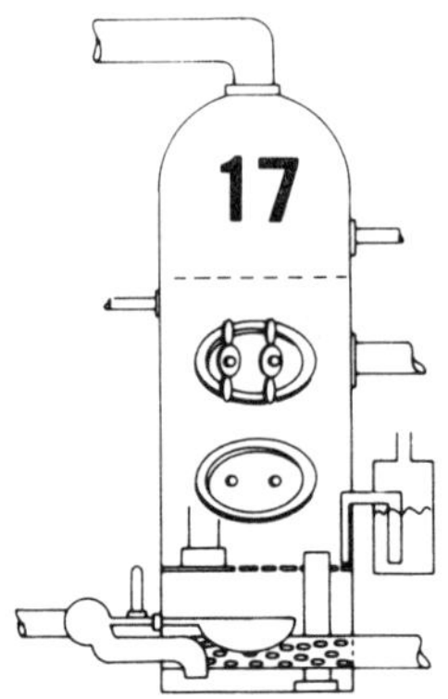

The Diesel Engine

A lot of people have tried to tell me that a diesel engine will run on alcohol. They usually obtained their information from someone such as a college professor with a Ph.D. in something totally unrelated to diesel engines and who probably figured it out with a pencil and paper.

By alcohol, they are of course referring to ethanol or methanol. As far as I know, no one has tried it on Vitamin A, also an alcohol. I am not saying that running a diesel on alcohol is impossible. It is simply that I have never seen anyone make it work. The reports I have read—such as masters theses from M.I.T.—all indicate that, if you can get one to run, severe hammering and knocking is the inevitable result.

This isn't meant to disparage those who figure things out with pencil and paper. Rudolf Diesel, the inventor of the diesel engine, spent a lot of time figuring out how things were supposed to work, using paper and pencil, before he started actual work on any of his engines.

His first engine was a closed-cycle steam engine using ammonia instead of water as the medium of heat transfer. He spent every spare franc (he was in France at the time) and every spare minute for several years trying to develop the engine. It never worked. However, the figures on paper indicated that it should.

On February 28, 1892, Rudolf Diesel applied at the German Patent Office for a patent on what we now know as the diesel engine. If some of the self-appointed experts would simply read what Diesel went through in developing his engine and deciding which fuels to use by personal experimentation, perhaps they wouldn't be so quick to pronounce the values of alcohol as a diesel fuel.

It is well-known in the 1980s that alcohol and gasoline are both excellent fuels for spark-ignition engines. It was also well-known before the 20th century and many pre-1900 automotive handbooks stated the fact. In the 1890s, hardly anyone knew what a diesel engine was, how it operated or what fuels were suitable for it. Diesel himself wasn't sure of what fuels to use. Once he tried gasoline and the resulting explosion blew an indicator off the engine. It narrowly missed Diesel and his assistant.

After acceptance tests in 1897, Diesel spent about two years trying to get the engine to run efficiently on coal dust. As late as 1940, one of his assistants was still trying it at another company. How the diesel engine works is as follows.

COMPRESSION

Air is compressed in the cylinder (Fig. 17-1) until it reaches a temperature of approximately 1000° F. The reason that diesel compression ratios are so high compared to gasoline engines is that every time the compression ratio of a piston to cylinder is doubled so is the amount of heat produced. This temperature is normally measured in degrees Kelvin, simply because on the Kelvin scale there are no "minus" temperatures and the math involved is much easier to work with. Try to figure out what 10 below zero is doubled and you'll see what I mean.

At this temperature almost anything with a "flash point" will ignite. Problems are created not by whether a fuel will ignite or not, but at the rate at which they expand. Gasoline and alcohol—once ignited—simply expand too rapidly. Both fuels would work fine if it weren't for all the parts connected to the piston that keep slowing things down. In an engine in which the piston was simply thrown away after each power stroke, hammering and knocking simply would never become problems.

There are such engines in common use today—those that throw the piston completely out of the cylinder at each power

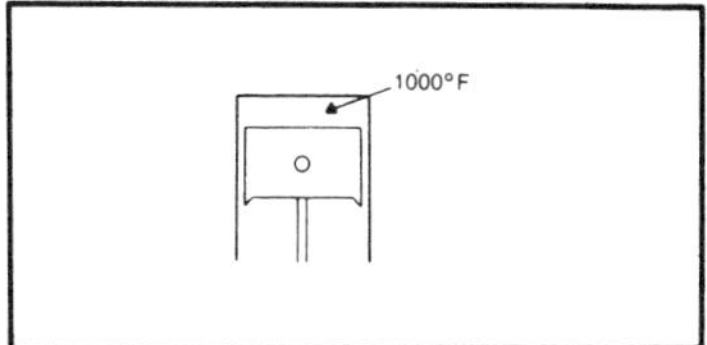

Fig. 17-1. Air is compressed in the cylinder.

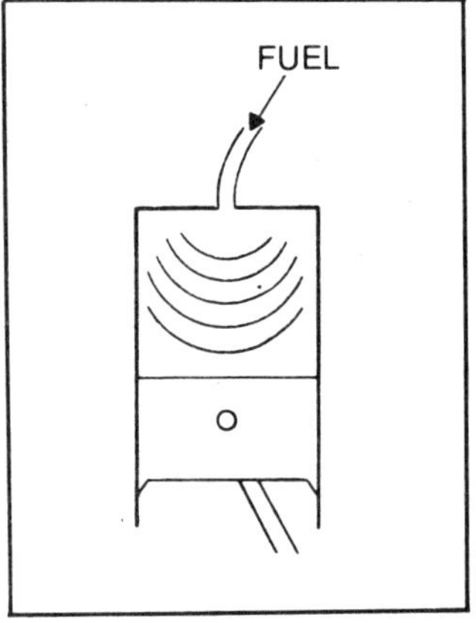

Fig. 17-2. Heat ignites the fuel and the piston is pushed down.

stroke—only they are known by another name: firearms. That is why some of our predecessors tried to run some of the first internal combustion engines on gunpowder. It makes almost as much sense as trying to run a diesel on alcohol.

At the end of the compression stroke, the fuel system injects the fuel. The 1000° F heat ignites the fuel which, in turn, expands to push the piston down to perform useful work (Fig. 17-2). How the fuel is injected under pressure and the equipment necessary to do so need not concern us here. There are already plenty of excellent books available covering diesel engine operation in more detail. I merely wanted to cover the basics before exploring the different types of fuels that will work in a diesel—such as vegetable oil.

CETANE RATING

Common diesel fuel has a *cetane number* just as gasoline or alcohol has an octane number. The cetane number is an indicator of how easily the fuel will ignite. A high cetane number means the fuel ignites readily at low temperature. A low cetane number means more heat is needed to ignite the fuel. A lower cetane number means the fuel is more likely to cause knocking and hammering. Alcohol sprayed into a cylinder will tend to accumulate and go off all at once. That is too much of an expansion at an instant. What is supposed to happen is the fuel is supposed to ignite and burn as soon as the injection spray starts. Consequently, there is an even pressure rise and no knock. It is a possibility that alcohol could be "doped" to act like diesel fuel, but it would probably be more trouble than it's worth.

VISCOSITY

In addition to cetane rating, there is *viscosity* to consider. Viscosity is the tendency of a liquid to resist flowing. Water has a low viscosity; it flows easily. Light oil is more viscous than water. However, it still flows easily. It too has low viscosity. Slow-flowing heavy oils have high viscosity. The reason low-viscosity fuels are used in a diesel is simply to enable them to flow easily through the fuel-pumping system. The viscosity cannot be too low or the moving parts in the pumping system will not be lubricated. If the fuel is too thick, the fuel will not spray into the cylinder easily or evenly and as a result will not burn well.

Viscosity is measured by simply timing the amount of fuel that flows through a certain-sized hole. The number of things—or fuels—that can be pushed through a hole in a fuel-pumping system and made to provide power in a diesel engine might surprise you. Everybody knows you can buy diesel fuel at the pump so let's dispense with that.

Many diesels can be run directly on crude oil. The sand and water in the oil can foul up the engine immediately if you pump the crude right out of the ground. Sand and water will all sink to the bottom of your barrel in a day or two, allowing you to drain the barrel from the top to use in your engine. The sulfur in the crude oil can't be filtered out this way, but it will take awhile to do the engine and pumping system any damage. The necessity of running a diesel on straight crude oil isn't something you are going to run into everyday, but it may be nice to know. You can never tell when you might get stuck in a diesel-powered tank next to an oil field in a war zone.

ALTERNATE FUELS

My own experience on running diesels on strange fuels is limited to using soybean oil. I had found a reference, to an M.I.T. thesis concerning running diesels on soybean and cottonseed oil, in a book on internal combustion engines. That's all it was—a reference. There was no how-to-do-it or anything.

Rather than guess at what I was doing, I called M.I.T. and ordered the thesis (#265499, Feb. 12, 1944) to read what had been done in the field before I dove off into it. There is much to be said for finding out what mistakes and failures somebody else had in the same area. You can read something a lot quicker than you can do it. Just make sure you get an account of someone who had "hands-on" experience and is not merely pontificating—or plagiarizing.

The thesis said simply to dump the soybean oil in and fire it up. Nothing could be that simple. It wasn't.

The first thing I did was reverse the polarity on the battery to the starter motor. The shop instructor told me I was lucky the battery hadn't blown up in my face. Great moments in science.

Once we got the diesel—a 600-horsepower Cummins turbocharged J-model—running on diesel fuel, we knew the only reason it would fail to run would be the soybean oil. We had eliminated all the variables but that one.

As soon as we knew the engine worked, I disconnected the fuel lines, drained the tank, and poured in a gallon of soybean oil I had bought at the local supermarket. This is expensive stuff, but it is still a lot better than having to walk if the need arose.

I hit the starter button and the engine roared to life. The smell of a diesel running on soybean oil is hard to miss. It smells like a hundred housewives burning supper. It also makes your eyes water.

Different vegetable oils do different things to your engine. Soybean oil actually produces about 10 percent more power than diesel fuel itself. Cottonseed oil produces about the same. Peanut oil gums up the works in the fuel-pumping system. What you have to watch with vegetable oils is that they start to solidify at various temperatures below 60° F. Throwing a gallon of soybean oil in a diesel-powered Oldsmobile in the dead of winter when you're in a hurry is simply not the way to do things. The vegetable oil will mix with diesel fuel (unlike non-anhydrous alcohol and water). Vegetable oil will not mix with gasoline.

At one time, vegetable oil was used extensively as a fuel for diesel engines in China—primarily tung oil. Tung oil is extracted from trees and is used in this country as a drying agent for varnishes. The Chinese went with tung oil simply because they had a lot of it and mineral oil deposits in China have always been extremely scarce or inaccessible.

The Chinese tried substitutes such as coal gas and producer gas. Neither was acceptable because both fuels cause heavy carbon deposits and wear in the engine cylinder. Other off-the-wall diesel fuels cause other problems. According to one story, Cummins Diesel ran some of their engines on buttermilk a few years back. Supposedly the fat in the milk caused the fuel system to gum up in five minutes or so.

One of my ambitions has been to run a diesel engine on fish oil. You could literally breed your own motor fuel. There might be a problem with the pumps getting gummed up with thick fish oil. Gumming the pumps would be an absolute with animal fat; but it would work.

Anything that has a flash point (minimum temperature at which oil will release combustible vapors) will work in a diesel. It is quite possible some genius will design a fuel-pumping system built on the principle of an electric toaster that will keep fat hot enough to flow uniformly into a cylinder head. What effect dumping tons of cholesterol into the atmosphere would have is anybody's guess. Possibly, pollution warnings would go from "smog alert" to "fat attack."

One additional note. A diesel can be run efficiently on alcohol if it is converted to spark ignition. But then it is no longer a diesel.

18 Steam Distillation of Vegetable Oils

The extraction of vegetable oil from grain crops is actually a much simpler process than extracting alcohol from the same crop. The primary difference is in percent yield. While a bushel of soybeans will yield 1½ gallons of soybean oil (or, if you prefer, diesel fuel), a bushel of corn will yield somewhere in the neighborhood of 2 pounds. However, let's leave the economics of it to others and merely concentrate on the how-to.

BASIC CHEMISTRY

Let's start with basic chemistry. Everything on earth has a boiling point—even iron. Heat iron enough and it becomes a liquid. Continue to raise the temperature and it becomes a gas. Various elements and compounds are labeled solids, liquids, and gases simply because that is how we identify them at the temperatures we are able to exist in.

Heat a soybean hot enough and it will become a liquid. Or parts of it will. If a soybean was 100 percent oil that simply needed to be heated up to liquify it, that would be the end of this chapter. All I would have to do is tell you what that temperature was and you could simply apply it and run your diesel.

Unfortunately, there are a lot of ingredients in a soybean that don't liquify until you get way past the temperature needed to liberate soybean oil. Try just applying heat and you will immediately have two problems:

☐ A destructively high temperature will be needed to extract very small amounts of oil.

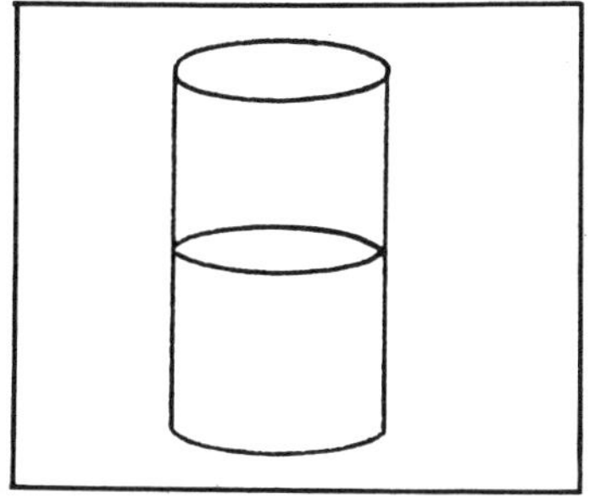

Fig. 18-1. A 55-gallon drum is half filled with water.

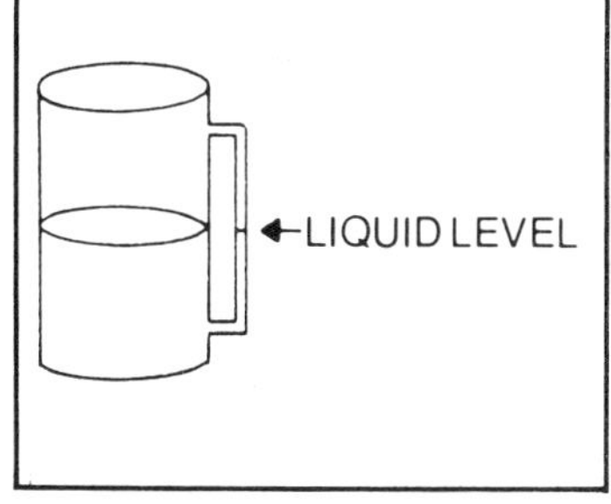

Fig. 18-2. A sight glass on the side of the drum reveals the liquid level.

☐ Mechanical entrapment of the oil by the other parts of the soybean will prevent complete removal.

The way these problems are avoided is to follow a few directions. First, grind your soybeans as finely as possible and dump them in water. If you want to calculate the percentage of oil yielded in this procedure, you will have to weigh both the soybeans and the water you intend to use before you start.

STEP-BY-STEP PROCEDURES

Start the project with a 55-gallon drum half-full of water. It's a simple concept (Fig. 18-1). Next, run a bushel of soybeans through a small hammermill, attrition mill, or roller mill and dump them in the drum.

To make life easier, you should install a sight glass on the side of the drum to tell you what the liquid level (Fig. 18-2) inside is at all times. It's not absolutely necessary.

The next step is to procure a lid for the drum and make two holes in it that will accept plumbing pipe and pipe connections(Fig. 18-3). The easiest way to mount this arrangement is to simply use pipe flanges of the same inside diameter of the holes, bolt them onto the drum lid (use some gasket sealer to prevent leaks), and

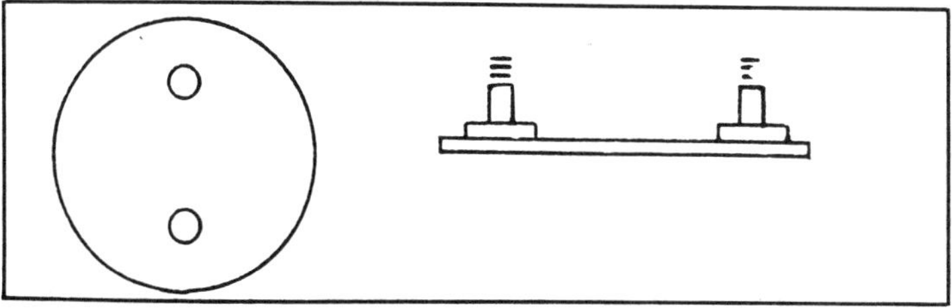

Fig. 18-3. A lid with pipe connections.

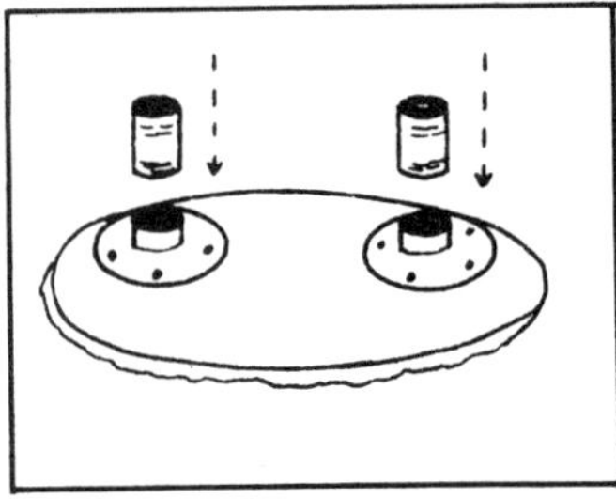

Fig. 18-4. The top of the lid subassembly.

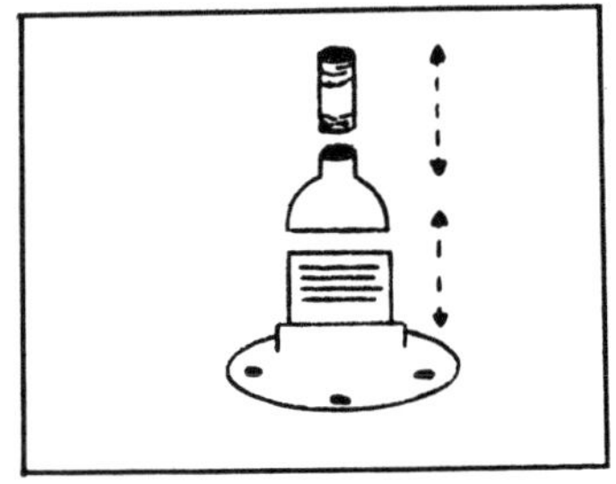

Fig. 18-5. Reducing couplings and pipe nipples.

screw two pipe nipples into the flanges. The top of the lid subassembly will look something like Fig. 18-4.

The inside diameter of the pipe flanges or pipe nipples is of importance only depending on how you want to continue your lines. Two reducing couplings mounted on the pipe nipples with a further pipe nipple is probably the easiest way (Fig. 18-5).

Attached to the uppermost (and smallest) pipe nipples will be lengths of heater hose with hose clamps to secure them tightly on the pipe nipples (Fig. 18-6). These hoses go off in two directions. One goes to a condenser and the other goes into what (in chemistry anyway) is referred to as a *dropping funnel*.

The easiest way to construct a condenser for this operation is to simply stick an old car radiator in a tub of water with heater hose attached. The whole setup will look something like Fig. 18-7.

Let's review part of what I have already described and continue with a full operation to produce 1½ gallons of diesel fuel from a bushel of soybeans. Fill the boiler one-quarter full with water. Observe it on the sight glass on the boiler. Grind up the soybeans as finely as possible. Dump them into the boiler. Use no more than a bushel. If soybean oil appears to be caking on your mill, you can—if you want maximum yield and have the room—dump the parts of the mill in the boiler.

Put the lid back on the 55-gallon drum boiler and seal it tightly to prevent air and steam leaks. Close the valve (use a globe valve

Fig. 18-6. Heater hose and clamps.

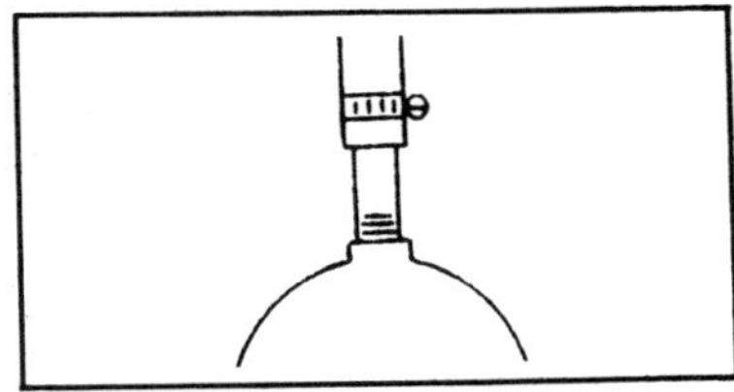

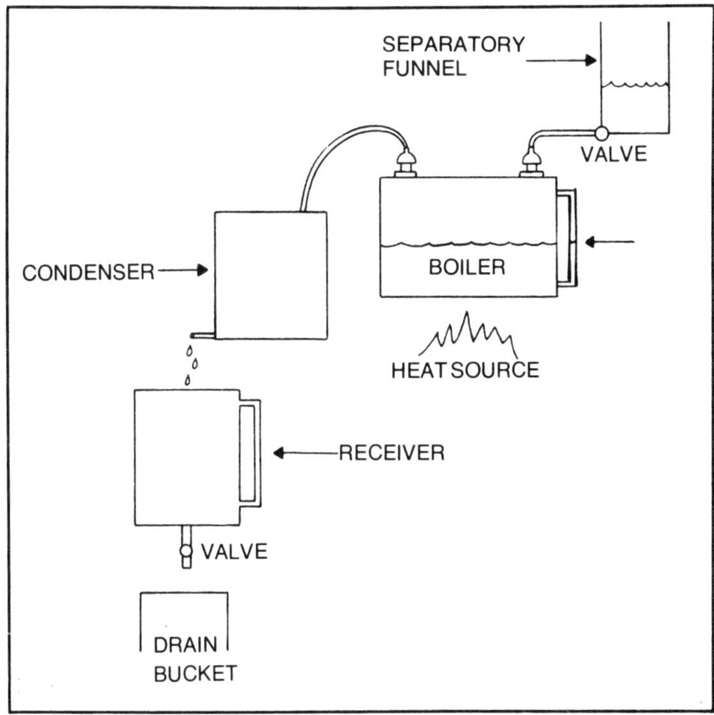

Fig. 18-7. Steam distillation setup.

with a handle) between the separatory funnel and the boiler. Fill the separatory funnel with water. A lid is not required on the funnel. Fire the boiler. While the boiler is heating to the boiling point, open the valve to the separatory funnel and fill the boiler to the halfway point.

The boiling point of the mixture is lower than the boiling point of either the water or the oil. What the temperature rises to is of little concern to us. If it boils, it rises and travels to the condenser anyway. To maintain an even boiling temperature, keep the water level in the boiler at the halfway mark by periodic additions from the separatory funnel. Flow rates of the condensor and the funnel should be identical.

Once the oil and water has traveled through the condensor, it is dumped into a receiver with a sight glass or a large glass bottle. The rest of the soybeans will stay behind in the boiler.

Add a few ounces of methylene chloride—obtainable from any chemical supply house—to the receiver. This merely helps to form a visible line between the soybean oil and the water in the sight glass or bottle.

Drain the receiver until the visible line disappears. Because oil floats on water, what you have done is just drain off the water.

To calculate the percentage yield of diesel fuel from soybeans, simply weigh the oil, divide by the weight of the soybeans, and multiply by 100. For example, if 10 pounds of oil is extracted from a bushel (56 pounds) of soybeans, the yield in percent is:

$$\frac{10}{56} \times 100 = 18\%$$

An external steam line—like that used for a stripper column in the steam chamber—can also be used. A trap must be included in case a drop in steam pressure causes the contents of the drum to be sucked back into the boiler.

19

Basic Chemistry

You've probably seen all the "chemical formulas" for alcohol by now that you care to. What I am attempting to do here is not simply display more formula, but instead teach you how to read the ones you have already seen. Simply knowing how to read the language of chemists opens up a lot of information that would otherwise be incomprehensible to you. On the other hand, if you've already had a couple of years of college chemistry, skip this chapter. You won't need it.

If you're like a lot of folks I've met you might wonder just what is meant by

$$C_2H_5OH$$

or

$$CH_3CH_2OH$$

and similar groups of letters and numbers. If the above two formulas confuse you a bit, don't worry about it. Both are "shorthand" forms of writing molecular diagrams known as *Lewis Structures*. The reason molecular diagrams are called Lewis Structures is simply because a chemist named G.N. Lewis hatched them in his clever brain.

ELEMENTS

The periodic table of elements is to a chemist what a slide rule used to be to an engineer. The periodic table lists in abbreviated form all the elements found in nature. That is, on the table, "N" stands for nitrogen. "O" stands for oxygen. "C" stands for carbon. "H" stands for hydrogen. And so on.

Therefore, $CH_3\ CH_2\ OH$ or $C_2\ H_5\ OH$ simply means that you are dealing with atoms of carbon, hydrogen, and oxygen. Now for the fun part the oil companies don't want you to know. The Lewis Structure for this molecule is:

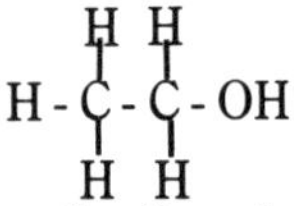

```
    H  H
    |  |
H - C - C - OH
    |  |
    H  H
```

This is ethanol. Notice the lines between the letters. Again, letters designate the type of atom. The lines stand for *covalent* ("sharing") bonds that hold the different atoms together to make up the molecule.

Here is another one that lots of folks claim is the "alcohol" that's used by Indianapolis 500 racers.

CH_3NO_2

or

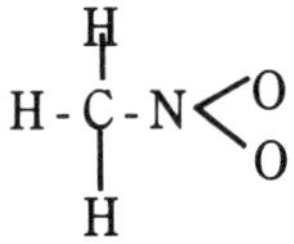

```
    H
    |
H - C - N < O
    |       O
    H
```

This is nitromethane. Technically it's not an alcohol; it's a nitro compound.

All of which leads us to the major difference between hydrocarbons and alcohols. The reason the oil companies don't want you to know the difference is twofold:

☐ There really isn't much difference. This is especially true for the "higher" alcohols.

☐ The oil companies produce most of our commercial alcohol—ethanol and methanol—themselves.

Let's start at the bottom. You should be familiar with methane gas. It is also known as swamp gas or sewer gas. Natural gas that you use to heat your home is 75 percent methane gas. Numerous books have been written on how to produce and use it. The Lewis Structure for methane gas looks like this:

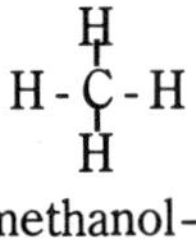

```
    H
    |
H - C - H
    |
    H
```

If you want to make methanol—or methyl alcohol—out of methane gas and water (on paper, anyway) you simply put methane gas and water together under heat and pressure:

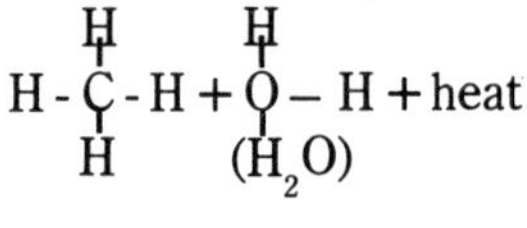

```
    H       H
    |       |
H - C - H + O – H + heat
    |
    H     (H2O)
```

What happens is that the heat breaks a hydrogen atom off the methane molecule and off the water molecule. You wind up with a methyl group and hydroxy group:

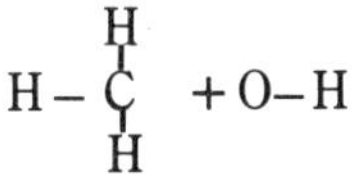

```
     H
     |
 H - C    + O - H
     |
     H
methyl group      hydroxy group
```

Anytime you have water and heat you will more than likely have steam—which is where the pressure comes from.

The pressure then forces the two disrupted molecules together and you wind up with methanol:

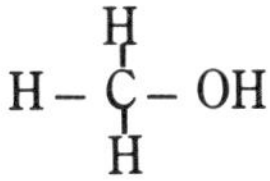

```
     H
     |
 H - C - OH
     |
     H
```

The oil companies use this same procedure to make ethanol out of ethane gas and water. Pick up a chemistry book and you'll almost always see *catalyst* mentioned for this type of reaction. All a catalyst does is speed up the reaction. This in turn conserves some of the energy required for the process. The reaction will proceed without a catalyst. Here is the difference between ethane gas and ethanol:

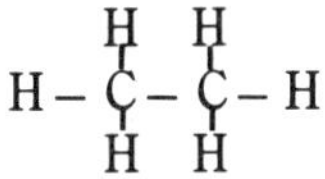

```
     H   H
     |   |
 H - C - C - H
     |   |
     H   H
   ethane gas
---------------
     H   H
     |   |
 H - C - C - OH
     |   |
     H   'H
    ethanol
```

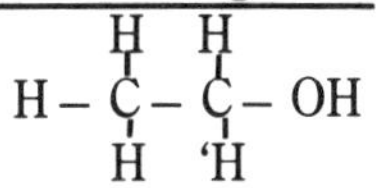

Not much, is there? Which poses another question; why is ethane gas a gas at room temperature and ethanol a liquid?

Ethane gas has an atomic weight of "30." That is, if you look at a periodic table of the elements you will see that carbon has an atomic weight of "12" and hydrogen is "1." The molecule has a total atomic weight of $2 \times 12 + 6 \times 1 = 30$. Oxygen, which has an atomic weight of 16, is always found in the atmosphere as a diatomic (two stuck together) molecule—atomic weight 32.

It would weigh exactly the same as methanol. So why does ethane gas and oxygen float at room temperature and methanol doesn't? There are two reasons for this difference.

The first reason is called *entropy*. Everything in the universe has a tendency towards randomness—including ethane gas. The second reason is called *hydrogen bonding*. Oxygen and hydrogen have a tendency to pull toward each other even when they aren't part of the same molecule. In the case of methanol, the oxygen of one molecule is attracted to the hydrogen of the next—sort of like this:

```
    H
    |
H - C - OH
    |
    H
    H
    |
H - C - OH
    |
    H
```

Multiply this by billions of molecules and you can see that the hydrogen bonding is stronger than entropy. You have seen this hydrogen bonding in your daily life. It causes "surface tension" on liquids and is why water beads and gasoline doesn't.

Obviously, the longer the chain of carbon atoms in an alcohol molecule the weaker the hydrogen bonding becomes because the percentage of oxygen becomes smaller. The smaller the percentage of oxygen in the alcohol the closer the properties of that alcohol to its corresponding hydrocarbon.

For example, there is a world of difference between methane and methanol. There is very little difference between octane (eight carbons with attached hydrogens) gasoline and octanol. Again, the percentage of oxygen in octanol is much lower. The oxygen in an alcohol does not provide energy (it has already been oxidized) and so must be counted simply as "dead weight."

That is why enlarging fuel passageways for metering alcohols, such as methanol and ethanol, in the carbureter are a must. The higher the alcohol the less enlargement you need.

With methanol—atomic weight total 32—half the weight is oxygen. The carburetor jet area (not diameter) must be made twice as large to meter sufficient fuel to the engine.

Look at butanol and see if you can figure out how much increased jet area you need:

```
    H   H   H   H
    |   |   |   |
H – C – C – C – C – OH
    |   |   |   |
    H   H   H   H
C – 12   H – 1   O – 16
```

If you figured out that 16/74 of this particular alcohol is not going to provide push to the pistons, you are correct. Converting to decimals gives you 22 percent. You need 22 percent more area for the fuel to pass through. Again, *not* diameter.

CARBURETOR JET SIZE

To convert 22 percent more area to jet size diameter— which is how jets are labeled—you will have to use algebra. In this particular case, you would work the problem as follows:

Let's say you had a two-barrel carburetor with two main jets of .050 each.

$\text{Area} = \pi r^2$ r in this case being equal to .025 (half the diameter).

$\text{Area} = (3.14)\,(.025)^2$

$\text{Area} = (3.14)\,(.0006)$

Area = .00188 square inches

To this figure we have to add 22 percent of itself:

$$\begin{array}{l} .00188 \\ \underline{.00041} \\ .00229 \text{ square inches} \end{array}$$

Now you have to convert this area figure back into diameter in order to ascertain the jet size(s) needed for butanol:

$.00229 = 3.14(x)^2$

$\text{Area} = \pi r^2$

Obviously "x" is the quantity we don't know.

To get rid of π— or 3.14 —we simply divide both sides by it:

$$\frac{.00229}{3.14} = \frac{3.14(x)^2}{3.14}$$

Which allows us to cancel out "3.14" on the right side of the equation:

$.0007 = x^2$

$.0265 = x$

If you wondered about my lightning-like calculations to find the square root of .0007 it was simple: I pushed a button on my calculator. Multiply .0265 by two to arrive back at a diameter. That is, your main jets should now be .053, or .003 oversize to accommodate butanol.

You can use this same system for other alcohols. Just don't take it as an absolute. The odds are the main jets in your carburetor weren't the right size to start with.

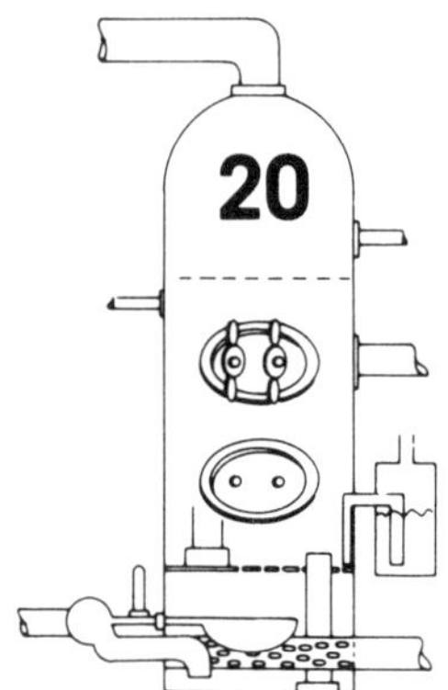

Basic Microbiology

Fermentation processes are controlled by microorganisms. The study of microorganisms is known as microbiology. If you already knew that, good for you. It took me over a year and a half of college chemistry before I discovered that you were supposed to go to a microbiologist for fermentation processes and to a chemical engineer for distillation processes. Hopefully, I can steer you in one of the proper directions by including the following textbook approach to the field of microbiology:

The word *microbiology* is a broad term meaning the study of living organisms that are individually too small to be seen with the naked eye. It includes the study of bacteria (bacteriology), viruses (virology), yeasts and molds (mycology), protozoa (protozoology), and some forms of life that do not fit well into any of these groups. Such minute forms of life are given the name of microorganisms.

Organisms are named using binomial nomenclature which consists of the genus and species. A common example used is man:

MAN

Kingdom—Amimalia (Metazoa)
Phylum—Chordata
Subphylum—Vertebrata
Class—Mammalia
Order—Primates
Family—Hominidae
Genus—Homo
Species—Sapiens

Thus, man is often referred to as *HOMO SAPIENS* using the binomial nomenclature system. Using this system, the genus and species are italicized. The genus is capitalized but not the species.

The organism which produces butanol from yeast is *Clostridium acetobulyliciun*. Because it belongs to the genus *Clostridium*, it has the ability to produce endospores. An endospore is a minute, highly durable body formed within the cell and is capable of developing into a new fully functioning organism. Endospores are non-active and they possess no metabolic activity of their own. When the spore comes in contact with the proper environmental conditions, the spore transforms itself back into a vegetative cell (Fig. 20-1). Therefore, this is not a reproductive process.

Clostridium Tetani and *Clostridium botulinum* are close cousins to *Clostridium acetobutylicium*, but where as *Clostridium Titani* produces tetanus and *Clostridium botulinum* produces botulism, *Clostridium acetobulylicium* is harmless. The production of the spore is the key reason for the disease production capability of *C Tetani* and it is why rusty nails are a good place to find this organism.

The most obvious property possessed by an endospore is its resistance to heat and chemicals. For example, an endospore might remain viable for many years—perhaps centuries—under conditions as they normally exist in the soil. Furthermore, while most vegetative cells are killed by temperatures beyond 60° to 70° C, endospores can survive in boiling water for an hour or more. There is no simple explanation for this resistance, but the fact that endospores contain very little free water is a major reason for this stability to heat.

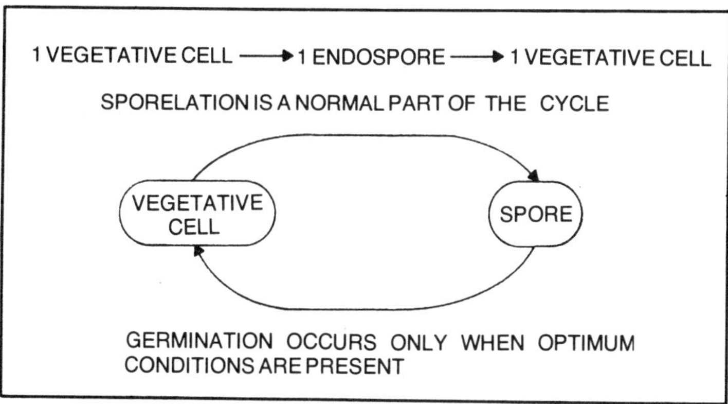

Fig. 20-1. Spore transformation.

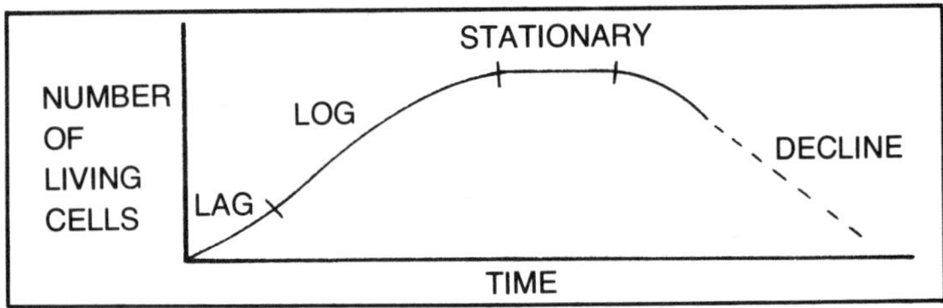

Fig. 20-2. Growth curve.

Most endospore forming bacteria are inhabitants of the soil, but bacterial endospores exist almost everywhere. The fact that endospores are so hard to destroy is the principle reason for the lengthy and elaborate sterilization procedures employed in hospitals, canneries and other places where absolute sterilization is required.

Bacteria grow geometrically (1,2,4,8,16 . . .). The growth can be represented by a curve. See Fig. 20-2. The growth curve consists of four stages: the lag phase, the log phase, the stationary phase, and the decline phase.

During the lag phase, the organisms are becoming adjusted to the new environment. The system is producing enzymes and other materials needed for growth. The system is metabolically active, but no reproduction is taking place.

During the log phase, the organisms are undergoing geometric division. The growth rate is at its maximum. There is a large increase in the number of cells and the rate of reproduction is much much greater than the rate of death.

During the stationary phase, the amount of nutrient per cell is decreasing as the population grows. Also, waste material is being excreted and it builds up. Conditions are unfavorable for growth and the growth rate slows down. During this phase, the rate of reproduction is equal to the rate of death.

During the decline phase, the culture medium is nearly exhausted and waste material is built up even more. Eventually, reproduction stops completely. In this period, the rate of reproduction is less than the rate of death.

CULTIVATION OF MICROORGANISMS

If you want to grow the critters, a semester worth of microbiology is included here as follows.

Nutritional Requirements

All biological systems share the same set of nutritional requirements.

Source Of Energy. All organisms require a source of energy. We can divide all living things into two broad categories based upon their principle source. *Chemotrophs* equals those organisms dependent upon the breakdown of chemical compounds for their energy. *Phototrophs* equals those organisms able to employ radiant energy.

Source Of Carbon. All organisms require carbon in some form. We can divide all living things into two broad categories based upon their principle source. *Autotrophs* equals those requiring only carbon dioxide with the ability to convert this to complex organic compounds. *Meterotrophs* equals organisms requiring an organic form of carbon.

Source Of Nitrogen. All living things require nitrogen in some form. Some bacteria are able to use atmospheric nitrogen and are referred to as *nitrogen-fixing bacteria* and are the only organisms capable of utilizing nitrogen in this form. Other bacteria are like plants and require nitrogen in the form of nitrates, nitrites, or ammonia. Still other bacteria are like animals and require nitrogen in the form of protein.

Source Of Sulfur And Phosphorus. All living things require sulfur and phosphorus in some form. Bacteria use phosphorus in the form of phosphate. The form of sulfur used depends upon the species.

Source Of Metallic Elements. All living things require numerous metals in various amounts depending upon the organism. These are Na, K, Ca, Mg, Mn, Fe, Zn, Cu, and Co.

Source Of Vitamins. A vitamin is defined as any essential factor required in small amounts for proper physiological maintenance. All living things require some vitamins in their diet. The number and amount of each depends upon the organism. Most bacteria are able to synthesize nearly all their vitamin requirements.

Source Of Water. All organisms require water.

The nutritional requirements for a given bacterium are reflected in the *medium* (*media* is plural) on which it is cultured. A medium of known chemical composition, both qualitatively and quantitiatively, is a *chemically defined* medium.

Physical Conditions

Temperature. Bacteria, as well as any organism, function best when cultured within their optimum temperature range as a consequence, they are classified as follows. *Psychophiles*: able to grow at temperatures between – 5 to 30 C; optimum range is 15 to

20 C. *Mesophiles*: able to grow at temperatures between 15 to 45 C; optimum range is 25 to 40 C. *Thermophiles*: able to grow at 45 to 75 C; optimum range is 60 to 70 C.

Oxygen. On the basis of their oxygen requirement, bacteria are classified as follows. *Obligate Aerobus*: those bacteria which will not grow unless oxygen is present. *Obligate Anaerobes*: those bacteria which can only grow in the absence of oxygen; the presence of any oxygen is lethal to them. *Facultative Anaerobes*: those able to grow in the presence or absence of oxygen; for some, the presence of oxygen results in better growth. *Microaerophiles*: those able to grow in the presence of only minute quantities of oxygen.

pH And Light. The optimum pH range is about 6.5 to 7.5. Light is necessary only for phototrophs.

Selective and Differential Media

An "all-purpose" medium is designed primarily to provide the nutrients necessary to support the growth of most microbes. For example, nutrient broth or agar, brain heart infusion broth or agar, tryptic soy broth or agar. An "enriched" medium contains additional supplements which have been added to enhance the growth of certain groups of organisms, especially those which are *fastidious*. Typical supplements are skim milk, albumin, sheep's blood, et al.

Selective media are enriched formulae that also contain a selective agent. Consequently, the growth of certain types of organisms is enhanced while the growth of other types is prevented. These media are especially useful for the specific isolation or cultivation of particular groups of organisms from mixed cultures. For example, sodium chloride agar contains 7.5 percent NaCl and is selective for *halophilic* organisms such as staphylococci. Sodium azide agar is selective for gram positives because the presence of sodium azide prevents the growth of gram negatives.

Differential media incorporate ingredients designed to cause certain organisms to develop a different appearance from others growing on the same medium. Therefore, it is possible to tell at a glance whether a certain colony is performing a particular chemical reaction or not. For example, blood agar, Simmon's citrate agar, and carbohydrate broth with phenol red are all considered differential media.

Selective and differential media combine the attributes of each type of medium. Thus the medium allows only a certain group of

Table 20-1. Carbohydrates Fermented by Species of Genus Clostridium Group II.

	12. C. ghoni	13. C. bifermenlans	14. C. Sordellii	15. C. lituseburense	16. C. limosum	17. C. subterminale	18. C. mangenolii	19. C. sporgenes	20. C. botulinum A, B, C, D, F	B, C, D, E, F	G	21. C plagarum	22. C. aclebutylicum	23. C. histolyticum	24. C. aurantibutyricum	25. C. novyi, Type A	Type B	Type C	26. C. perfringens	27. C. haemolyticum	28. C. felsineum	29. C. chauvoei	30. C. seplicum	31. C. defficile
Amygdalin	–	–	–	–	–	–	–	–	–	d	–	–	d	–	–	–	–	–	–	–	d	–	–	–
Arabinose	–	–	d	–	–	–	–	–	–	d	–	–	+	–	+	d	–	–	–	–	+	–	–	d
Cellobiose	–	–	–	–	–	–	–	–	–	–	–	–	–	–	–	–	–	–	–	–	+	d	–	–
Dulcitol	–	–	–	–	–	–	–	–	–	–	–	+	+	–	+	–	–	–	d	–	+	d	+	d
Eseulin	–	–	–	–	–	–	–	–	–	–	–	–	d	–	–	–	–	–	–	–	d	–	d	–
Fructose	–	d	d	+	–	–	–	+	+	+	–	+	+	–	+	d	–	–	+	+	+	+	+	+
Galactose	–	–	–	–	–	–	–	–	–	d	–	+	+	–	+	–	–	–	+	–	+	+	d	–
Glucose -	w	+	+	+	–	–	–	+	+	+	–	+	+	–	+	+	+	w	+	+	+	+	+	+
Glycerol	–	d	d	–	–	–	–	d	d	d	–	d	–	–	–	d	–	–	d	d	–	–	–	–
Glycogen	–	–	–	–	–	–	–	–	–	–	–	+	+	–	+	–	–	–	d	–	d	–	–	–
Inositol	–	–	–	–	–	–	–	–	–	d	–	–	d	–	–	d	+	w	+	d	–	–	–	d
Inulin	–	–	–	–	–	–	–	–	–	–	–	–	d	–	–	–	–	–	d	–	d	–	–	–
Lactose	–	–	–	–	–	–	–	–	–	–	–	+	+	–	+	–	–	–	+	–	+	+	+	–
Maltose	w	+	+	+	–	–	–	d	d	d	–	+	–	–	+	d	+	–	+	–	d	+	+	–
Mannitol	–	–	–	–	–	–	–	–	–	–	–	–	d	–	–	–	–	–	–	–	–	–	–	+
Mannose	–	d	–	d	–	–	–	–	–	+	–	+	+	–	+	–	–	–	+	–	+	+	+	+
Melezitose	–	–	–	–	–	–	–	–	–	–	–	–	–	–	–	–	–	–	–	d	–	–	–	d
Melibiose	–	–	–	–	–	–	–	–	d	–	–	–	–	–	–	–	–	–	–	–	–	–	–	–
Raffinose	–	–	d	–	–	–	–	–	–	–	–	–	d	–	+	d	–	–	d	–	d	–	–	–
Rhamnose	–	–	–	–	–	–	–	d	–	–	–	–	d	–	–	–	–	–	–	–	+	–	–	–
Ribose	–	d	d	d	–	–	–	–	–	d	–	–	–	–	–	d	d	w	d	–	–	+	d	d
Salcin	–	–	–	–	–	–	–	d	d	–	–	–	+	–	+	–	–	–	d	–	+	–	d	d
Sorbitol	–	d	–	–	–	–	–	d	–	d	–	–	d	–	–	–	–	–	–	–	–	–	–	d
Sorbose	–	–	–	–	–	–	–	–	–	–	–	–	d	–	+	–	–	–	–	–	–	–	–	–
Starch	–	–	–	–	–	–	–	–	–	–	–	d	+	–	+	–	–	–	+	–	d	–	–	–
Sucrose	–	–	–	+	–	–	–	d	d	d	–	+	+	–	+	d	–	–	+	–	+	+	–	–
Trehalose	–	–	–	–	–	–	+	d	–	d	–	–	+	–	–	–	–	–	d	–	–	–	d	d
Xylose	–	d	d	–	–	–	–	–	–	–	–	–	+	–	+	–	–	–	–	–	+	–	–	+

R.E. Buchanan & N.E. Gibbons, co-editors, Bergey's Manual of Determinative Bacteriology (8th ed., Baltimore: The Williams & Wilkins Company, 1974), p. 558.

organisms to grow (selective) and distinguishes between various types of reactions within that group (differential). Mannitol salt agar is an example.

POTENTIAL

To find out what a microorganism is capable of, consult *Bergey's Manual*. Any college science department should have one lurking around someplace. *Bergey's* will tell you at a glance whether a microorganism will do something or whether it won't.

See Table 20-1. A "-" in the column means it won't do something and a "+" means it will. For example, let's take our butanol-producing organism Colstridium acetobutylicun and see what it will do for us. If you read Table 20-1 correctly, what you get is that it—the C acetbutylicun—won't do anything with cellulose, but it will ferment almost any sugar or starch. That is, you don't need enzymes to convert starch.

"a.ce.to.bu.ty'li.cum. English n. acetone; M.L. adj. butylicum butylic; M.L. neut.adj. acetobutylicum referring to production of acetone and butyl alcohol.

"Straight rod 0.6-0.9 by 2.4-4.7 um; often granulose positive. Motile with peritrichous flagella. Spores are oval, subterminal; no exosporium, no appendages. Gram-positive. Cell wall contains DL-diaminopimelic acid.

"Surface colonies are circular, 3-5 mm in diameter, raised, irregular margin, grayish white, translucent, glossy surface.

"Little growth in nutrient broth. Abundant growth in nutrient broth containing fermentable carbohydrate.

"Fermentation products include acetic and butyric acids and butanol. Acetone produced.

"Glucose gelatin hydrolyzed; milk coagulated, stormy fermentation; milk digester slowly by some stains.

"Fixes atmospheric nitrogen.

"Optimum temperature for growth is 37 C.

"Has been found in soil.

"The G + C content of the DNA is 28-29 moles %.

"Reference strain: ATCC 824, NCIB 8052."*

---

*R.E. Buchanan & N.E. Gibbons, co-editors, *Bergey's Manual of Determinative Bacteriology* (8th ed.; Baltimore: The Williams & Wilkins Company, 1974), p. 561.

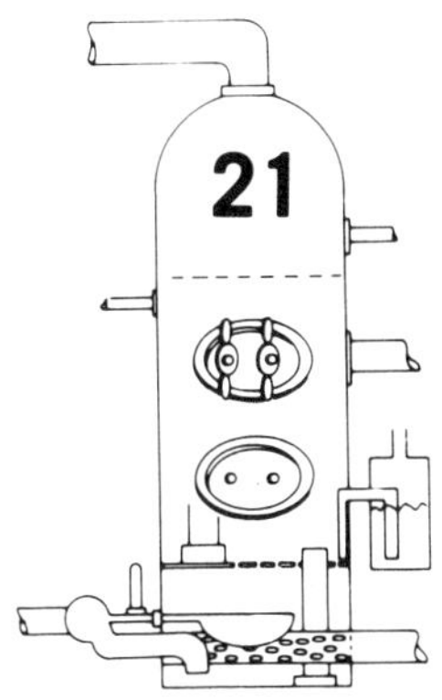

Basic Jet Sizing

The drilling out of the carburetor main jet or jets to accommodate alcohol is a fairly common practice now. There is no question that it works. The problem is that it never works as well as it should. There are couple a problems with such a crude method that surface every now and then.

ENLARGEMENTS

The first is that, without machine tools and the equipment to hold the jet, there is the distinct possibility that the hole you are trying to enlarge will wind up being aimed in a different direction than it was originally intended—sort of at an angle. (Fig. 21-1).

In fuel injection equipment this is critical. With some carburetors, such as the Tillotson found on a Harley-Davidson, fuel sprayed along the venturi tube walls instead of down the throat due to misdirected fuel orifices will cause the engine to quit almost altogether. It won't do the system any good, regardless.

The second problem that sometimes occurs is that of the drill gounging up a sliver of brass inside the fuel orifice of the jet. This can cause the same type of flow that you get when you stick your thumb part of the way under a water faucet. The jet size you thought you drilled will not be the one you are operating on. If you look at a normal main jet, you will see recessed openings (bottom of Fig. 21-2) that were put there to prevent such accidental gouging in handling and shipping.

Another problem that I haven't seen addressed anywhere is that of the proof strength of the alcohol causing problems in engine

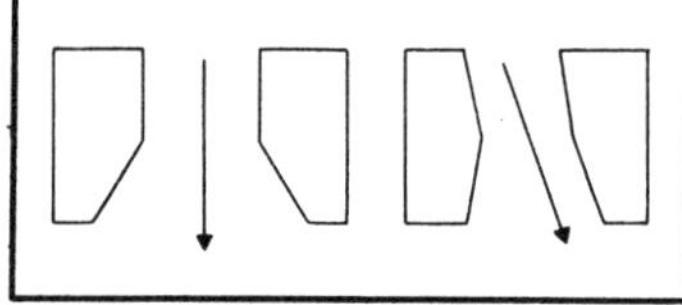

Fig. 21-1. Recessed openings on a carburetor main jet.

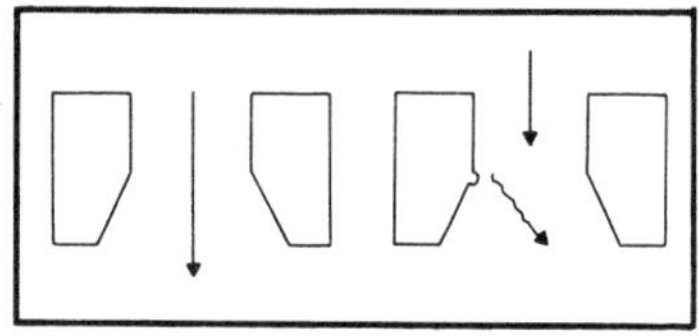

Fig. 21-2. The carburetor main jet on the right has been gouged.

operation. One common complaint seems to be that, once the proof of the alcohol gets below 160 or 165, the engine misses and sputters. It shouldn't.

My 10-year-old son (this was in 1979), had no problems at all running a mini-bike on 140-proof rubbing alcohol. However, he couldn't get it to idle. We knew that the main jet couldn't be at fault because we had a main jet adjustment. We couldn't adjust the idle circuit. Therefore, it had to be the fuel. That is what you will find if you just apply a little common sense and basic chemistry to it.

First, water won't burn; at least not in or at the normal temperatures and pressures utilized in the standard internal combustion engine. In case this seems like a strange statement, ask around. You would be amazed at the number of people who think you can save motor fuel by "adding water."

Second, the percentage of water in the alcohol cuts down the amount of alcohol itself that can flow through a main jet. Silly statement, isn't it? Let's see what happens when Joe Mechanic gets his engine running on alcohol and then starts lowering the proof.

The first thing Joe does is enlarge the hole size of the main jet to accommodate the oxidized oxygen stuck on the enthanol molecules. See Fig. 21-3. What he just did was enlarge the hole

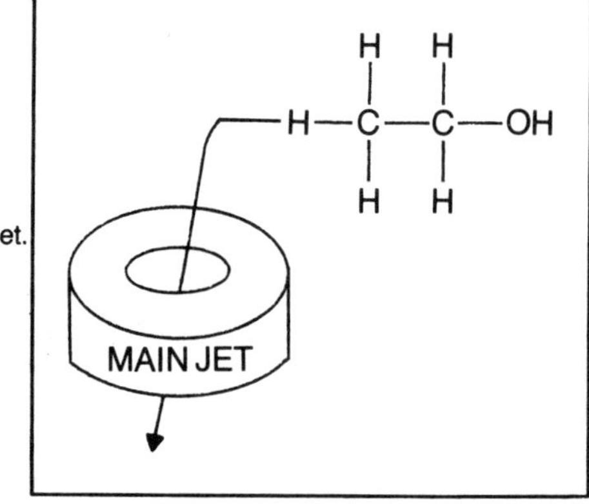

Fig. 21-3. An enlarged main jet.

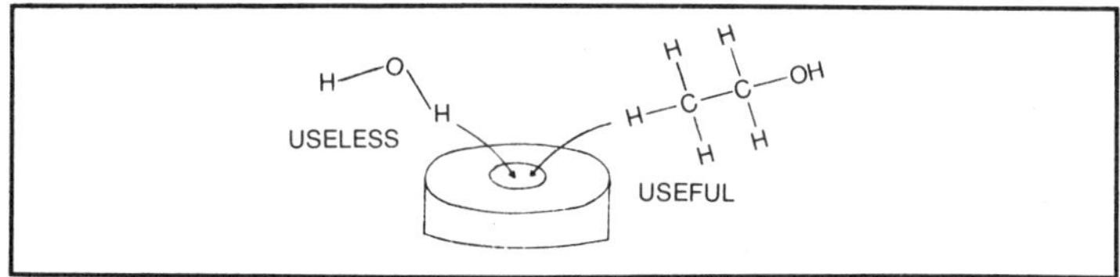

Fig. 21-4. Water molecules take up space and the alcohol molecules ignite.

size 30 percent in area—not diameter—to accommodate a fluid that is 30 percent dead weight. At 200 proof—100 percent alcohol, no water—everything works just peachy.

Of course, it wouldn't work at all if he hadn't allowed for the 30 percent dead weight. However, at 160-proof he starts to run into problems. The engine misses and dies under load. The conclusion he reaches is that 160 proof alcohol simply doesn't get the job done. It's not the alcohol. It's the main jet again.

Here's what happened. Joe Mechanic now has two separate types of molecules flowing through his main jet. *Alcohol molecules* ignite in the combustion chamber and provide power and *water molecules* simply take up space. See Fig. 21-4.

At 160-proof, 20 percent by volume of the fuel is taken up by non-performing water molecules. But he didn't enlarge the main jet 20 percent in area, did he?

The sputtering and missing are simply clues that the engine is starved for fuel. It would have done exactly the same thing with a 10 percent surface area increase in the main jet opening if Joe had switched from gasoline to alcohol with the same method.

The idling in such a situation is even worse because the standard idle circuit opening is 1.2 to 1.5 times the size of that of the main jet. It's the same problem only proportionately larger.

ALTERNATIVES

In a pinch, there are a couple of other methods that can be used that don't involve changing the main jets. If you can't enlarge the

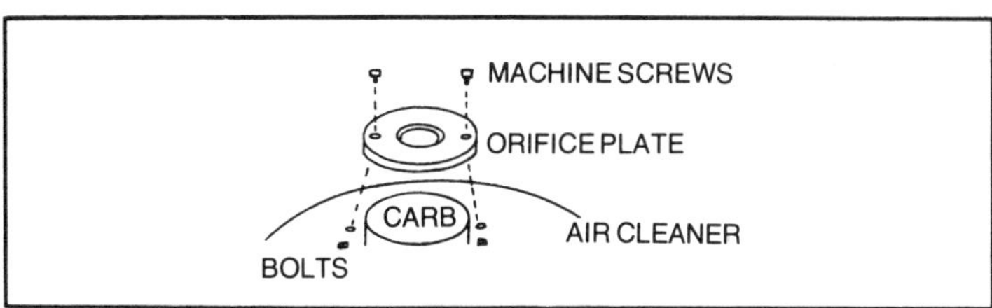

Fig. 21-5. An orifice plate is installed on top of the cleaner.

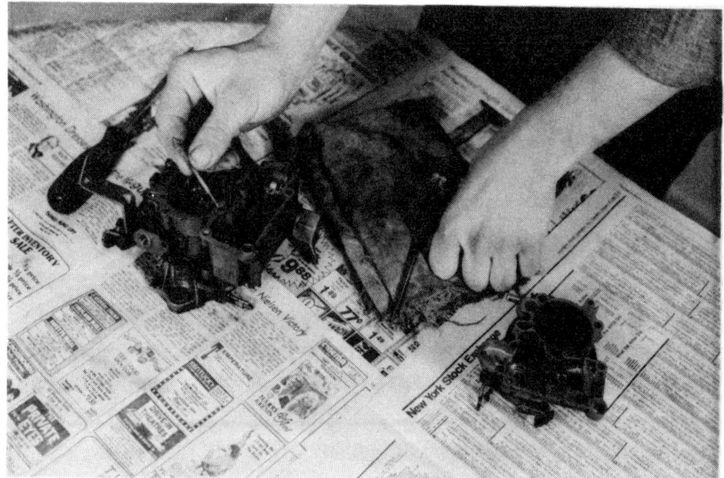

Fig. 21-6. A typical 6-cylinder carburetor with removable needle valves.

fuel hole, reduce the size of the air opening. The first method of reducing the air opening is to simply partially close a manual choke. It's not the best way to do things, but it works. The second method is to simply install an orifice plate (Fig. 21-5) on top of the air cleaner. This isn't the greatest way to do things either because it reduces the difference between the main throttle bore and the venturi tube. Again, it works.

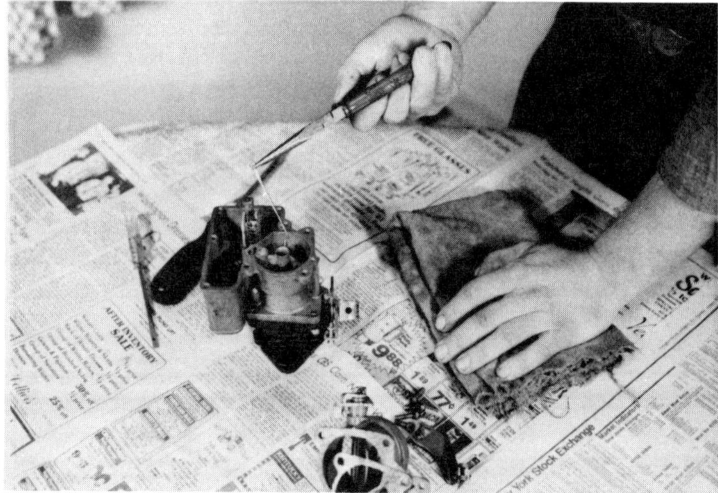

Fig. 21-7. Basic jet sizing.

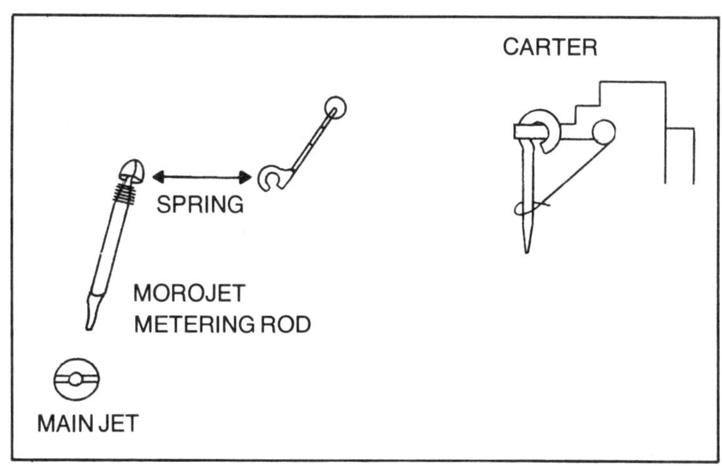

Fig. 21-8. The metering rod goes through the main jet.

There have been carburetors with variable venturis made in the past that would also work, but it's unlikely you'll find one in the local junkyard. They're pretty rare. There are a few carburetors that do not require either replacement or modification of the main jet. Both are used on 6-cylinder engines.

The single-barrel Carter and the Rochester Monojet have metering rods that go down through the main jet. The idea of a metering rod, tapered at one end, is to vary the main jet size under varying speed and load conditions for better fuel economy. See Figs. 21-6, 21-7 and 21-8. If you remove the metering rod, held on in both cases by a spring, what you have just done is enlarge the flow area of the main jet enough to accommodate ethanol.

On a 1969 Chevy Nova with a Monojet carburetor, we went from 23 mpg on gas, down to 19 mpg on ethanol, and then back up to 22 mpg on ethanol after we put a few miles on the old clunker. The car had over 100,000 miles on it when we got it. Apparently, the alcohol burned all the carbon off the heads and in turn relieved engine friction.

The beauty of the metering-rod removal on a single barrel is that—with a little practice—you can effect changeovers in either direction in less than five minutes. That is, in clear, warm, and sunny weather.

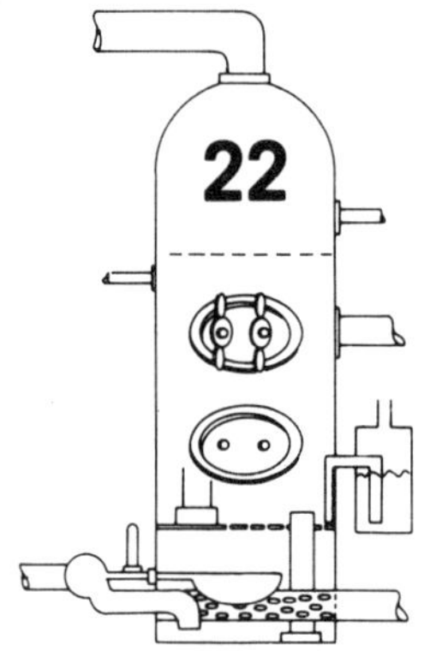

22 Matching Carburetors With Engines

One of the most common mistakes made in building either single-fuel or dual-fuel alcohol/gasoline carburetion systems seems to be that when people change carburetors they don't pay attention to what the carburetor was designed for and what size engine each carburetor is intended to fit.

One such example—a rather extreme one—happened to me in the summer of 1979. Almost no one knew anything about the conversion of standard gasoline carburetors to run alcohol. I was awakened by a phone call at 1:00 in the morning by an individual from a "research center" who was almost in hysterics. I had been coaching him and his associates by telephone for several weeks and I had been able to get them through most of the basics. They had a copy of the first book I wrote on the subject, but they were apparently unwilling to take the time to read it—hence all the telephone coaching.

The problem was that they had installed a carburetor with an adjustable needle valve main jet adjustment on an engine and couldn't get the engine to fire on either gasoline or alcohol. According to them, the resulting embarrassment in front of the news media they had invited was all my fault. It was a type of carburetor I had recommended.

The type of carburetor they had failed with was from a Harley-Davidson motorcycle, 74 cubic inch. The engine they had put it on was a 250-cubic inch truck engine. My wife had heard the entire conversation and, at the end of it, rolled over and said: "Good grief, even I knew better than that!"

A motorcycle carburetor belongs on a motorcycle—not on a truck. At the risk of sounding a little oversimplified, the truck engine couldn't breathe. It was starved for air and fuel. Obvious, isn't it?

Where most mistakes are made by people with "hands-on" experience is in the opposite direction—too much carburetor for the engine. Such an arrangement will often get you from point A to point B, but it will never run completely right. There will be "flat spots" on starting off that can simply never be eliminated. Such a "flat spot" is caused by to much air entering the engine at low rpm. Flat spots will also occur under acceleration. This latter problem can be cured by enlarging or lengthening the accelerator pump circuit. That is, if you don't mind squirting half a pint of fuel into your engine to accommodate your oversize carburetor every time you want to pass someone on the interstate.

CALCULATING CARBURETOR FIT

There is a way to compute carburetor fit without having to bolt everything up only to find out that it simply won't work the way you want it to. First, carburetors are rated according to the amount of cubic feet of air they will allow into the engine in a given 60-second period. A 650 CFM (cubic feet per minute) double-pumper Holley is rated to allow 650 cubic feet of air per minute into the engine. Believe me when I tell you it also lets in plenty of air at low rpm. The amount of fuel allowed into the engine is determined by jet size and engine rpm.

Second, engines are rated by the volume of each cylinder at bottom dead center of piston stroke multiplied by the number of cylinders in the engine. If each cylinder at bottom dead center measures 30 cubic inches and there are 6 cylinders, then you have a 180-cubic inch engine, or 6 x 30. It is simply the procedure of matching these items up that allows you to discover whether a certain carburetor will work on a certain engine the way it is supposed to.

Obviously, you can't match a 650-CFM carburetor with a 400-cubic inch engine unless you convert cubic feet to cubic inches, or vice versa, in order to have a common denominator. Let's say that you want to install a Holley double-pumper 650-CFM carburetor on a 400-cubic inch 1972 Ford (Fig. 22-1). Will it work right or won't it?

First, convert the engine cubic inches to allow for the fact that air only enters the cylinders on one of the four strokes (that of intake). No air from the 650-CFM Holley enters the engine during compression, ignition or exhaust. Therefore, you have to divide the number of engine cubic inches by four. Only 100 cubic inches of air enter the engine per revolution.

Second, figure out the rpm's the vehicle will be operating at in order to determine the amount of air the engine requires. To do this, look in a spec book such as *Motor's* or *Chilton's*.

What you will find is that horsepower for a 1972 Ford at 4000 rpm is 172. At 4000 rpm, the engine is wound all the way out and the vehicle is traveling approximately 100 miles an hour. If you were building a stock car, you could do the calculations at this point because 100 miles an hour is a valid and reasonable figure for stock cars. However, this is a super no-no on the interstate.

The speed at which the engine was designed for maximum power is that indicated by torque at rpm. In this case, it is 298 foot-pounds at 2200 rpm. The rpm figure is the one with which to work. You must allow for not carrying a load and operating at half-throttle. The idle speed will be in the 500 to 625 rpm range.

Multiply 650 CFM by 12 cubed, 12 x 12 x 12, to get cubic inches in place of cubic feet—1728 x 650 (or 1,123,200 cubic inches). That is, the carburetor will deliver over one million cubic inches of air to the engine in 60 seconds.

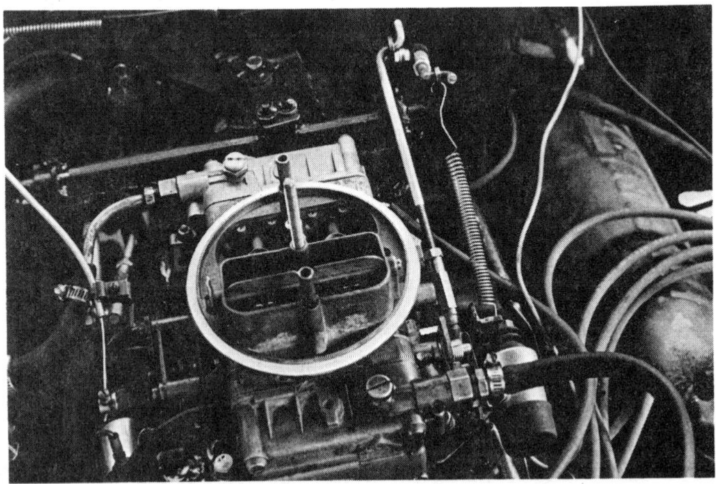

Fig. 22-1. The front two barrels and float chamber meter alcohol to the engine. The back section is for gasoline. Only one fuel can be used at a time. This unit was a prototype on a 1972 Ford truck owned by the author. Similar units are available from: AFI, Box 302, North Dayton Station, Dayton, OH 45404.

Multiply the amount of cubic inches—100—by the rpm at maximum torque, 2200. What you get is 220,000. You have a carburetor that will deliver over a million cubic inches of air sitting on top of an engine that requires less than half that much air at either full load or speed. At this point you could legitimately surmise that you have a bit too much carburetor for the engine.

Fortunately for us, the air entrance into the engine is controlled somewhat more by piston suction than it is by the size of the hole the air passes through. If it weren't, the engine wouldn't work at all. What does happen is that at low rpm's the air openings in this particular carburetor are so large and they allow so much excess air into the engine that it requires intense concentration on the part of the operator to nurse his machinery along until he reaches highway speed. Engine spitting and coughing becomes quite common.

In the process of switching carburetors and engines, you might run across a few that are ideal for using with different fuels that don't seem to make a whole lot of sense. One such carburetor I found a few years ago was titled "Winton" (not to be confused with the entire automobile of the same name, no relation) and had every adjustment conceivable. These particular carburetors were used for race cars during the 1930s. What made this particular carburetor so appealing to me was that in the 1930 nobody took a race car to the track on a trailer. The car was driven to the track on gasoline. At the track, the gas tank was emptied and filled with grain alcohol, the carburetor was "dialed in" in a few laps around the track on alcohol and the car was ready to race.

At some point, someone would pull a spark plug, see that it wasn't fouled, and pronounce the car ready to run. A lot of "how-to-run alcohol" books pontificate on reading spark plugs, but I'll pass. It's an art in itself. Of the 30 or so people employed by the spark plug companies to "read" spark plugs for competition engines, only about three of them actually know what they're doing. The other 27 are simply giving bad advice that results in blown engines.

The "Winton" puzzled me for days. I couldn't figure out how to mount it until I found out that it had come off a car built in the 1920s.

It was an updraft. I had been trying to stick it on upside down. I hope you won't have the same problem when you come across exotic-looking adjustable carburetors. Just keep in mind what to look for and why.

CARBURETOR CLASSIFICATIONS

Practically everyone is familiar with the standard downdraft carburetor in widespread use since the 1930s. Prior to the 1930s, practically all automobile carburetors were updrafts. In some specialized applications, such as motorcycle or marine, side-draft carburetors are still used.

Rather than assume that everyone else knows more than I did when I started, I'll try to give brief illustrations and examples of each. The updraft was the first carburetor ever used to any large extent. On the old in-line engines, it was mounted on the bottom of the intake manifold and sucked air (later mixed with fuel) at a right angle and then up ("updraft") into the engine. This was ideal for a gravity-fed fuel system (Fig. 22-2).

You might think that the advent of the V-8 engine and the fuel pump eliminated this design in passenger cars by the 1930s, but that's not how it happened. Somebody finally figured out—after millions of updraft carburetors had been manufactured—that one basic premise had been overlooked. The premise that updrafts were built on was that gasoline vapor is lighter than air and rises.

It isn't and doesn't. It's heavier than air and falls. Hence the downdraft carburetor. In all fairness to the ancient updraft, it still has a couple of uses. You will find a lot of them on farm tractors complete with main jet adjustments. A flick of the wrist is all it takes to modify them to run on alcohol. You will also find them on the bottom of some small airplane engines. It makes it easier to look over the engine and see where you're going.

The *sidedraft*—also known as a crossdraft, horizontal, or side-outlet carburetor—is most often seen on motorcycles. Harley-Davidson has been using sidedraft carburetors on their

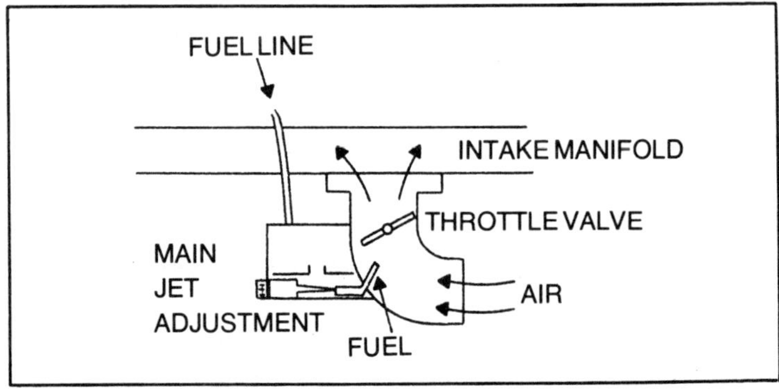

Fig. 22-2. The updraft carburetor.

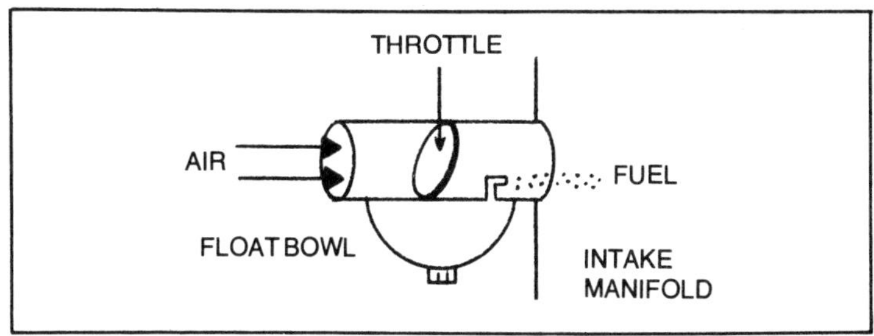

Fig. 22-3. The sidedraft carburetor.

v-twins since they began building them at the turn of the 20th century. The primary advantage of the sidedraft is that it eliminates one right-angle turn the fuel and air has to travel through. Sidedrafts (Fig. 22-3) are also used on stationary power plants and European cars.

If you're going to play with sidedrafts and updrafts (many are immediately adjustable for alcohol) you had better stick to in-line engines such as sixes and fours. The intake manifolds of such engines are easy to modify and there is plenty of room under the hood. The basics of matching carburetors with engines can be summarized as follows:

☐ Will the configuration fit? If you try to put an updraft on a V-8, the answer is obviously no.

☐ Is it the right size? Size can be determined by a specification chart or book, by comparing engine sizes—a carburetor off a 472-cubic inch Lincoln should work equally well on a 445-cubic inch Oldsmobile—or by trial and error. The last method is not advisable unless there is no other information available.

☐ What was the carburetor designed for? A carburetor for a stationary power plant might get you from point A to point B used on a car but you're not going to pass anyone without an accelerator pump.

The best part of all this is that once you learn how to switch carburetors around you can do what lots of other folks are doing. They are boxing carburetors and selling them as "alcohol carburetors" to the unsuspecting public.

Just be sure that you use fiberglass fuel tanks. If you don't, you'll wind up with cheesy-looking stuff clogging your carburetor, pressure regulator and fuel lines. This might cause you to think that you made a mismatch when the entire problem was just dirt.

Designing Dual-Fuel Units

There has been a ton of literature printed on how to convert gasoline carburetors to alcohol. The technology for such conversions is so well-known that it does not need to be rehashed here. There are a couple of serious drawbacks to such a one-fuel conversion.

The first draw back is fuel availability. There simply are not enough people producing alcohol in enough different locations to insure that the motorist is going to be able to refuel with anything resembling convenience. If you are really desperate, you can run gasoline through a carburetor modified to run alcohol; but the amount of fuel wasted is horrendous. The last time I tried it, I managed to drive a 6-cylinder Chevy Nova 180 miles on 40 gallons of gas.

The second drawback—even if you can find the stuff—is the price. To my knowledge, as of this writing, no person or group of persons has made a serious attempt to compete with gasoline on a nickel-for-nickel basis. Everyone I've run into so far wants to charge one-third more for a fuel that will only propel a vehicle two-thirds as a far as a gallon of gas will.

Compression ratios can be raised to equal the efficiency of gasoline, but then you can no longer run gasoline and the expense is great enough to pay for a two-year supply of gasoline anyway. Apparently, one of the main reasons fermentation alcohol is so expensive from individuals is the common practice of simply throwing the byproduct away. A gallon of alcohol leaves over 6 pounds of byproduct. At current prices, this byproduct is worth 7 cents a pound and up.

Of course, you can always use methanol. It is still relatively cheap. Methanol does have a few drawbacks such as toxicity. If you drink just a taste of it, it can blind or kill you. Even the fumes are toxic. Catch a whiff of methanol fumes and what you get is a headache. If you're lucky. Can you imagine a million methanol-powered cars on the Los Angeles freeway and all pumping out toxic fumes? All the aspirin in the world wouldn't cure the resulting headaches and it would take the entire United States Army several weeks to carry away the bodies.

Methanol also causes "puddling" in the intake manifold of a low-compression engine. It is much harder to start in cold weather than ethanol. And it takes a lot more methanol than it does ethanol to get from point A to point B.

About the only way I've been able to solve the supply and price problems of ethanol as a motor fuel is to make my own alcohol and then run it in a carburetion system set up to run on alcohol or gasoline. This is similar—in principle if not in mechanical design—to the add-on propane units currently in use. One popular magazine touted a "dual-fuel" unit that involved taking the carburetor apart every time you wanted to change fuels. Of course, you could always carry a spare carburetor. Either way the requirement of changing jets or carburetors out in the middle of nowhere, in pitch black night, and in subzero weather could get on your nerves. It would mine.

Obviously, the primary reason for building a dual-fuel system (Figs. 23-1 and 23-2) is convenience. If you're driving along and run out of alcohol and have to pull into a gas station, you want to be able to merely flip a switch or lever to change fuels. There are a number of ways to approach the problem.

COMMON DESIGNS

The first and most obvious solution is to simply install two carburetors. Shut one off while the other one runs and vice versa. It's not as easy as it reads. A lot of patents have been applied for and granted on systems that simply won't work.

One of the most common designs is that of a carburetor stuck at each end of the intake manifold. Even if you could get this setup to work at all, what would happen is that a carburetor providing fuel at one end of an engine would provide an over-rich mixture at the cylinders next to it and an extremely lean mixture at the cylinders farthest away. Provided you could get such a system to keep an engine running for any period of time, what would happen is that

Fig. 23-1. This is the first alcohol/gasoline dual-fuel unit built in the United States. It was built for AFI of Dayton, Ohio by Kinsler Engineering of Troy, Michigan in September, 1979. The tube coming out of the intake manifold is a heat riser from exhaust going to a baffle box underneath the carburetors.

the near cylinders would become clogged with carbon and the far cylinders would wind up with burned valves.

Once you understand that both carburetors, especially when operating at different times, have to pass fuel and air through the same opening in the intake manifold (preferably the one cut in at the factory that built your vehicle), you will have to decide how to shut one carburetor off while you activate the other one.

Fig. 23-2. The AFI dual-fuel unit. The throttle linkage and the air cleaner are shown.

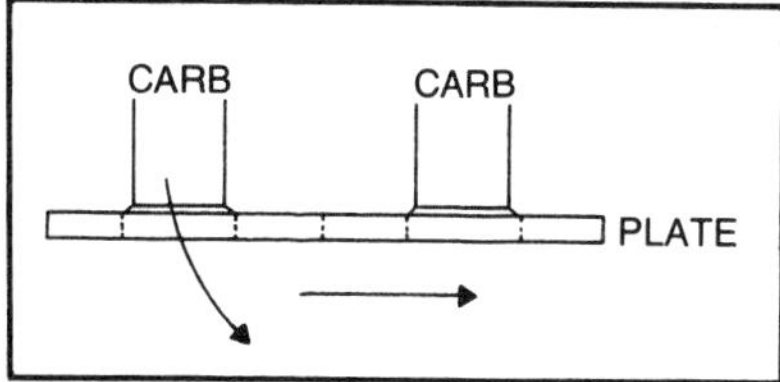

Fig. 23-3. A sliding plate type of valve.

The most common method is simply to install a valve or series of valves at the base of both carburetors. The closing of such a valve stops the suction created by the pistons when the engine is turning over. Consequently, no gasoline or alcohol is sucked out of its respective float bowl.

Certain types of valves won't work for very long. One is a sliding plate type of valve with holes offset in it to block and open the venturi passageways (Fig. 23-3). In an indoor shop kept at constant temperature, this would work. On the road, temperature variations between the engine block and the outside atmosphere would cause it to warp and jam.

Another system that might work is a rotary valve. A cylinder with holes in it is placed under the carburetor with holes drilled in such a fashion that rotating the cylinder opens one fuel passageway or the other. You can't afford leaks around the cylinder. The minute you get a speck of dirt in the assembly you have just converted back to a single fuel system.

The ideal way to do it is to install butterfly valves (or flipper valves) underneath the carburetors. One valve opens as the other closes. If these valves are not leveled exactly like the carburetors themselves, they get hammered out of shape by constant usage and cause fuel and air leaks (Fig. 23-4).

The only serious problem I encountered with such a unit—other than the flipper valves being hammered out of shape in about six months, making it impossible for the system to idle on alcohol—was the fuel recondensing during the trip through the fuel

Fig. 23-4. Valves.

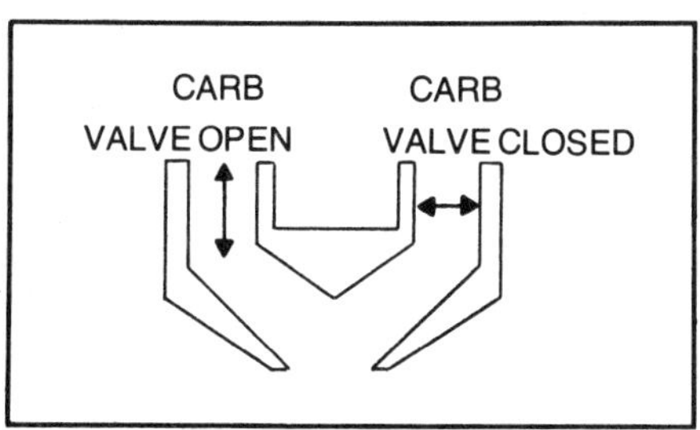

passageways. Consequently, a lot of liquid fuel that should have been a vapor was pumped through the engine and out the tailpipe unused. This sort of thing is especially noticeable in cold weather. On a cold, dry night, running on alcohol looked as if I was laying down a smokescreen for a regimental-sized infantry attack.

Utilizing one carburetor is a much better system. The linkage problems are more complex. On a two-carburetor system, you simply hook the butterfly (in carburetors) valves up in series. That is, both throttle valves in the carbs open and close at the same rate—controlled by accelerator pedal linkage. Because the fuel and air flow to one of the carbs is shut off, what its throttle does is of no consequence.

BUILDING A SYSTEM

Building a dual-fuel carburetor system requires the observation of certain chemical principles and mechanical designs. If you don't allow for each and every detail of the following description, something is not going to work.

The first factor is that alcohol must have a larger hole to flow through. You must either be able to vary the size of the hole(s) you are using or provide an alternate set of holes somewhere else. This latter approach is used with the Holley double-pumper carburetor because it has float chambers at both ends. The secondary side of the Holley has to have an accelerator and idle circuit installed. That is the easy part. The hard part is in designing the linkage to control the two pairs of butterfly valves in the carburetor itself.

The second rule of chemistry you must learn is that alcohol, other than anhydrous, simply will not mix with gasoline. The water in the alcohol causes the two fuels to separate with alcohol on the bottom and gasoline on the top. If you have both fuels in a float chamber, the alcohol will the the first fuel to enter the engine. As a result on a cold morning, the fuel simply won't ignite. The two fuels must be kept separate.

Once you have figured out how to keep the fuels from mixing, there are three circuits in the carburetor you have to concern yourself with:

—Idle circuit.

—Main circuit.

—Accelerator circuit.

The idle circuit drips fuel into the engine when the throttle valve is completely closed. The fuel must be extremely rich (gasoline about six parts air to one part fuel) to keep the engine

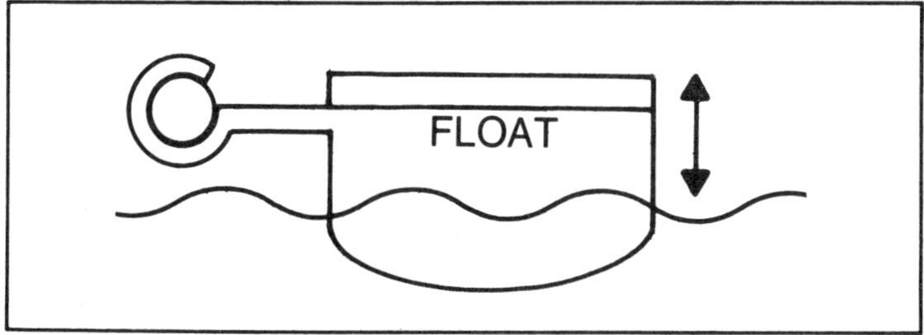

Fig. 23-5. An open-barrel carburetor.

running at low speeds. At low rpm some of the plugs will not fire. It's even worse with alcohol.

If you don't have a Holley double-pumper or two-carb system and want to be able to sit in traffic running alcohol without feathering the gas pedal, you will have to install an extra plate setup for idling only underneath the carburetor. Any idle system installed above the throttle valve simply will not work. If it does and the throttle valve is open, it is not an idle circuit.

The main circuit is simply the main jet or jets. You must either be able to modify the size of the existing jets or install an external main circuit and float chamber. My own design for this is simply an iron ring placed inside the air cleaner with a fuel line emptying over the center of the carburetor. An electric fuel pump can be used to pump the alcohol to the float chamber. It is cheaper—but somewhat more complicated—to use directional valves to route the fuels through a conventional fuel pump (Fig. 23-5).

The iron ring merely stabilizes the fuel line and keeps it from flopping around inside the air cleaner. The float chamber simply meters the alcohol. Figure 23-5 represents a one-barrel carburetor. You will need a fuel discharge nozzle for each venturi tube on the primary side of your carburetor. I would appreciate a phone call before you begin mass-producing these things; this one is patent applied for.

The third mechanical marvel to duplicate in your system is an accelerator pump. Like the idle circuit, it is possible to get from point A to point B without it. What will happen is if you "goose it" your engine will simply quit. Driving gingerly from place to place is fine. It's just that if you run into an unfriendly tractor-trailer bearing down on you broadside, you will have to get out of its way in the same fashion. Gingerly.

You would probably be better off installing an external accelerator circuit. All an accelerator pump does is squirt in an extra "shot" of fuel so that the engine speed doesn't have to increase in order to suck more fuel out of the float bowl. Throttle response is immediate.

Normally, an accelerator pump is simply a springloaded plunger. Slow movement of the plunger up and down in a cylinder filled with fuel doesn't do anything. A rapid movement of the plunger downward creates enough pressure to force fuel up a passageway into the engine. This is what happens when you "tromp on it." The spring merely returns the plunger to its original position. See Fig. 23-6.

Stick these three items together and you have a fuel system (or carburetor). It doesn't matter where you put the linkage as long as it is controlled by the same linkage that controls your original carburetor. There have been systems patented that use two foot pedals, but the amount of linkage involved is a nightmare.

If you use electric fuel pumps in both fuel tanks, you can use one float chamber for both fuels. Just be sure that one fuel is completely gone before you start pumping in the other.

Two fuel tanks (Figs. 23-7 and 23-8) are essential because the fuels are not miscible. What I normally use in designing a system in place of an auxiliary tank and fuel pump is a plastic milk carton downstream of the fuel pump filled with whatever I'm trying to use as fuel. A mechanical pump will suck fuel out of a milk carton just a quickly as it will a conventional gas tank. Just be sure the hose from the milk carton to the inlet side of the fuel pump is placed all the way in the bottom of the milk carton. The milk carton is also a cheap way to check mileage.

Linkage to switch fuels can be mechanical, electronic or hydraulic and describing them would require a book in itself. Use your imagination. Some of you will probably come up with things that no one ever dreamed of. Just be sure you have calculated the proper size holes for all three circuits (a longer stroke can be used

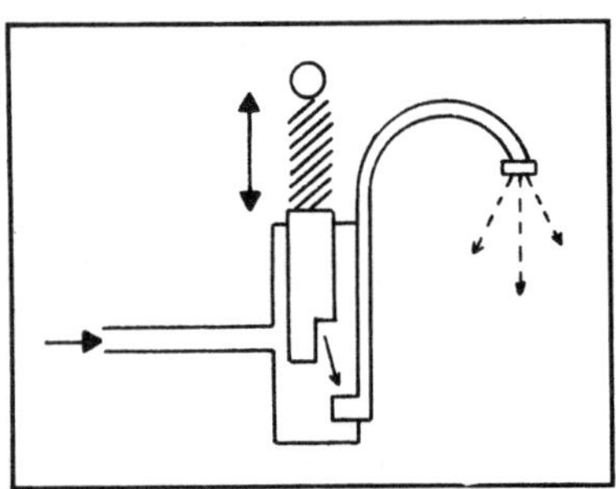

Fig. 23-6. The spring returns the plunger to its original position.

Fig. 23-7. The tank on the right of this 1972 Ford truck is for alcohol.

on the accelerator circuit) and test them to see if each circuit *works* before you go hogwild on the linkage.

THE FLOAT CHAMBER

The float chamber does not have to be attached to the rest of the carburetor. It's simply a convenient way to do things if you happen to be casting several thousand of them at a time. It also takes up less room.

Fig. 23-8. The push-pull lever operates the linkage to a Holley 4-barrel. The two switches are for the electric fuel pump in the alcohol tank and to the solenoids on the side of the carburetor to lock open the idle circuit being used.

For a prime example of a float chamber external to the machinery, observe a toilet tank. It works exactly the same as a carburetor float with one minor difference. The carburetor float is emptied by piston suction while the toilet tank is emptied by atmospheric pressure (after, of course, you hit the handle).

In experimenting with different fuels or carburetion systems that the carburetor was not designed for, you occasionally encounter problems with the float chamber. Before describing such problems, I will outline a basic float chamber to remove any doubts about how one works. Don't feel insulted if you already know this, just skip a few paragraphs. Readers who don't know will appreciate your tolerance.

To begin with, a float chamber merely keeps the fuel at a certain level. If the fuel level is too low, the engine will suck the fuel out faster than it can be replaced—causing the engine to starve. As you can see in Fig. 23-9, the minute the thin layer of fuel in the bottom of the float chamber is reduced any further, what the engine will be sucking through the main jet is air.

This problem can be cured by simply setting the float higher in the float chamber. Many carburetors have "inspection covers." This is simply a threaded screw or bolt on the outside of the float chamber for the purpose of inspecting the liquid level. The idea is that if liquid—more than a drop or two—slops out the hole your float is set too high. If nothing comes out it is set too low.

If it weren't for the fuel pump constantly pushing in fresh fuel, setting the float too high wouldn't be of concern. You wouldn't even need a float. The way a float works is as follows. Fuel comes through an orifice at the top of the float chamber. The float is pushed upward by its own bouyancy and the pressure of the liquid beneath. Finally, the fuel underneath pushes the float up to where it blocks the fuel orifice and stops more fuel from coming in. As the engine sucks fuel out the float drops. More fuel comes in. And so on.

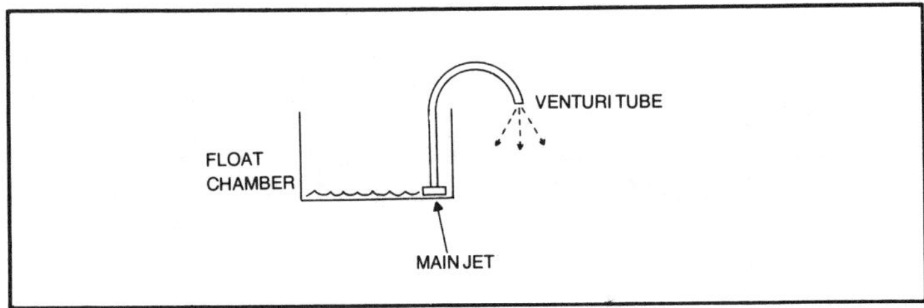

Fig. 23-9. The float chamber keeps the fuel at a certain level.

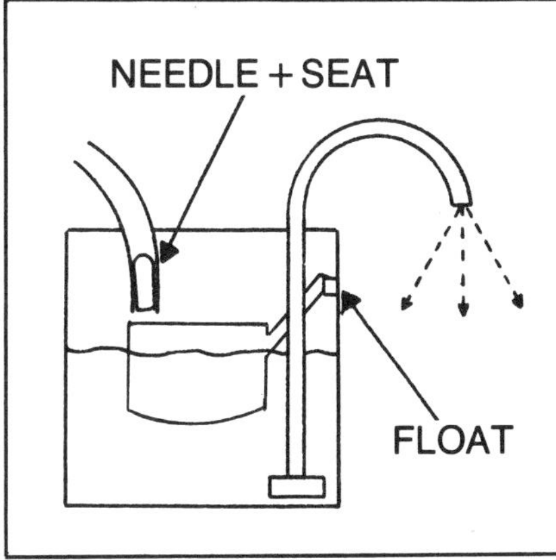

Fig. 23-10. A side view of a float chamber.

To keep the system operating without leaks, instead of the float merely blocking a hole, the float pushes up a piston inside a cylinder. The piston either blocks or admits fuel. This is called a needle and seat. Opposite the needle and seat, the float will swing on a hinge that allows it to move up and down (as opposed to side to side or all over the place). If you go to a junkyard and buy an old carburetor to play with, this section will make a lot more sense to you. A side view will look something like Fig. 23-10.

The needle will usually be tapered in such a fashion that the further the float sinks the wider the fuel passageway becomes. The pointed end always goes to the top. In dealing with alternate fuel and carburetion systems this is where many problems originate. If you build a venturi system that requires less fuel, what happens quite often is the fuel pump pushes in fuel at a higher pressure than the float is capable of withstanding. The fuel will then be pushed out the atmospheric vent.

The same thing will happen if you use an electric fuel pump. Some of the them will behave themselves for a few days and then wet all over the place. Either reason for overflowing can be cured by installing a pressure regulator between the fuel pump and the float bowl. The amount of pressure regulated depends on so many variables (strength of fuel pump, fuel, engine size, carburetor size, and so on) that you will simply have to experiment. Approximately 2 to 3½ pounds is average.

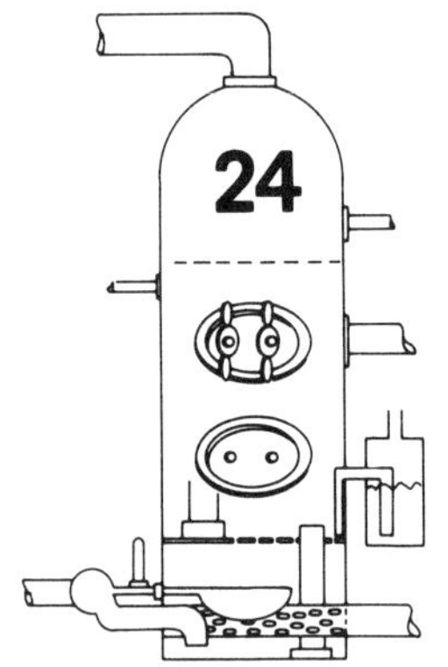

24 Heat Risers and Preheaters

At the turn of the century, a lot of cars presented cold weather starting and operating problems. If you could get a 1909 Winton to start in subzero weather, you couldn't get it to run. And if you could get it to run, sometimes ice would form in the carburetor venturi tube and choke off your air supply.

There were two cures for these problems. In order to keep an engine running in severely cold weather, the vaporized fuel had to be hot enough to ignite. The higher the compression, the hotter the fuel/air mixture got. In those days, a high-compression engine was one that had a ratio of 4½ to 1. The low-compression problem was solved in 1926 with the addition of tetraethyl lead to gasoline. This allowed compression ratios to be raised past 7 to 1. In turn, this created sufficient heat in the engine to allow it to start and keep running in severe weather.

A similar situation exists today in alcohol-powered engines. An engine with an 8 to 1 compression ratio simply does not create enough heat to fully utilize a fuel that is capable of performing smoothly at ratios in excess of 14 to 1. Unless you want to spend hundreds of dollars reworking your engine, you will just have to live with it. However, there are a couple of other ways to impart heat to the fuel and air. You heat the alcohol to keep it from thickening. At 60° F, ethanol is the consistency of molasses and simply won't work. You heat the air to keep the carburetor throttle valve from icing over.

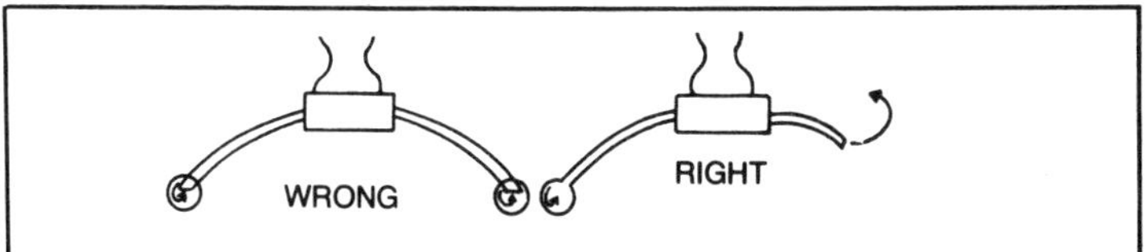

Fig. 24-1. Exhaust gas flow.

SCAVENGING HEAT

There are three areas you can scavenge heat from. One is the electrical system. The voltage regulator shuts off the generator (or alternator) periodically in order to keep the excess electrical power generated from blowing up your battery. Scavenge this electrical energy and you have a source of heat. Obviously, this can get pretty complicated and it is the least practical way to do things. It is also the most expensive.

The two most common inexpensive ways of doing things are to scavenge either exhaust heat or use hot water from the cooling system. A word of caution if you use exhaust heat. If you play into an exhaust pipe you *must* vent the exit end to the open atmosphere. If you don't, the pressure on each side of the heat riser tube (or line) will be equal in both directions. As a result, there simply will be no flow of hot exhaust gas through the line. See Fig. 24-1.

The two places you can route exhaust heat are between the base of the carburetor and the intake manifold or up inside the air cleaner. Don't try to warm the fuel with exhaust heat or you will develop a severe case of vapor lock. If you're lucky. If you wrap hot metal around a fuel line and develop a leak anywhere in the system, what you have just created is a fire and explosion.

To run exhaust heat to the base of an alcohol carburetor, construct a box with entrance and exist lines for the exhaust line. Put a baffle plate with small holes on each side of the box far enough apart to allow for a cylinder between them to accommodate the air/fuel mixture from the carburetor throat. Use holes slightly large enough to create more surface area than the exhaust line has. What will happen is that at low speeds the hot exhaust gas will simply pass through and not warm much of anything—preventing vapor lock. At high speeds, when a tremendous amount of cold air rushes into the engine, the speed of the exhaust gas causes the baffle plate chamber to absorb heat before the exhaust can escape. This in turn helps to transfer heat from the baffle plate chamber to the incoming air/fuel mixture (Fig. 24-2). There are dozens of variations of this design. This one explanation should be enough to

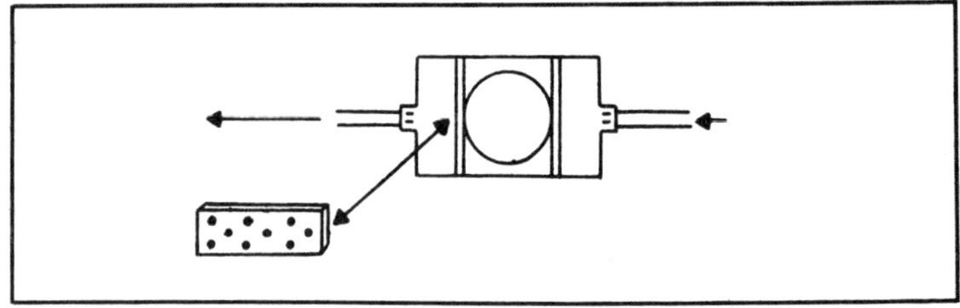

Fig. 24-2. Heat transfer.

get you started on building and designing adapter-plate style heat risers.

The other area to route exhaust heat is inside the air cleaner. If you run a tube of material that does not conduct heat very well from the exhaust manifold into the air cleaner, change the material to copper (an excellent heat conductor), coil the copper flat inside the air cleaner, and then route it back out of the air cleaner. You should be able to eliminate any carburetor icing problems that develop. See Fig. 24-3.

A similar procedure can be used with hot water from the heater hoses if you find that you are getting too much heat. What happens here is that the cold air flowing over the copper coil absorbs heat and carries it (the heat) past the throttle valve and into the engine.

Hot water from the heater hose can also be used to warm up the fuel line leading to the float bowl. One popular magazine recommended wrapping a fuel line around a radiator hose, but that is ridiculous. The thermostat has to open before any hot water whatsoever is allowed into the radiator via the top radiator hose. The thermostat doesn't open until the engine is at full operating temperature. That is a little late to get a preheater to work.

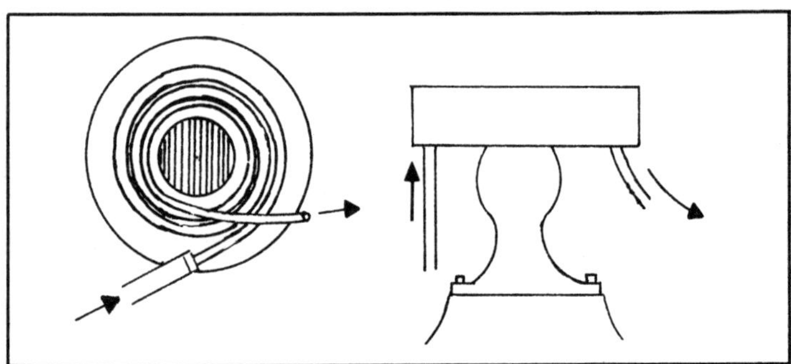

Fig. 24-3. Use copper coil to eliminate carburetor icing.

CARBURETOR ICING

In addition to failing to provide enough heat to ignite alcohol properly, there is another problem you might encounter—*carburetor icing*. There are two main reasons for carburetor icing. First, with a partly closed throttle, the pressure drop across the throttle opening generates a temperature drop. If the atmospheric humidity is high, ice will sometimes partially clog the throttle opening. When this happens, the throttle opening has to be increased to allow for the ice that has just formed. If this ice-formation increased-throttle opening game of musical chairs keeps up for any length of time the only immediate way to cure the problem would be with a hammer and an ice pick to chip away the ice. This problem can sometimes be prevented by the addition of water-soluble antifreeze to the fuel. But only if the fuel discharge is at the throttle edge.

Second, the evaporation of fuel can cause the same problem. Ice has a tendency to form where fuel spray hits an unheated surface. This used to be common in aircraft carburetors. Originally, they tried to solve the problem by heating the intake air. This worked, it was inexpensive and yet at the same time caused some loss of power. And sometimes it caused detonation. If you run out of patience, you can stick an aircraft carburetor on your car. Because an aircraft carburetor automatically provides for changes of altitude (and temperature) it is ideal for someone living in Denver.

Most of the time, carburetor icing takes place between 28° F and 55° F in atmospheric conditions of 65 percent to 100 percent humidity. Icing usually occurs after the engine has been started and before it reaches operating temperature. This is especially prevalent in alcohol-powered engines because alcohol burns cooler to start with. The exhaust heat from an engine running on alcohol will often be 200° F cooler than that of the same engine running on gasoline.

If your engine stops or stalls at idling speed, then icing might be your problem. That is provided you reworked your idle circuit properly. The symptoms are almost identical.

Ice formation on a throttle plate will look something like Fig. 24-4. The ice has a tendency to form on the edge of the throttle plate. You normally don't have these problems with a standard automobile running on gasoline for a couple of basic reasons:

☐ That's what the system was designed for. The temperature and pressures the unit was designed to operate at were figured

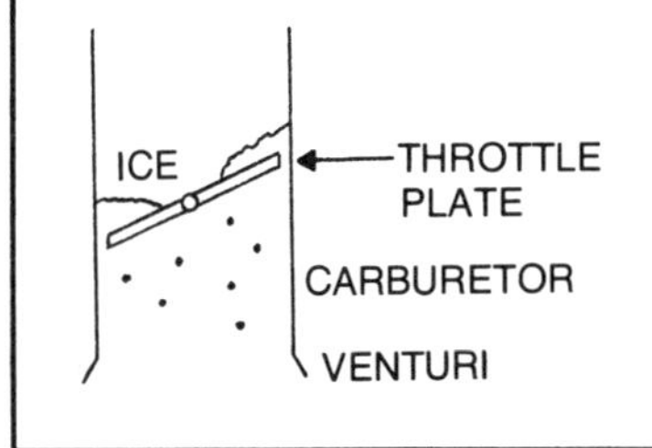

Fig. 24-4. Ice formation on a throttle plate.

out on paper by some of the top engineers and mathematicians in the country.

☐ Even after all the top brains have figured out everything on paper and failed to make it work, they still have access to tons of diagnostic equipment and large chunks of money to find out "why not?" Normally, Detroit doesn't put things on the market unless most of the bugs have been ironed out. This ironing out process has so far taken over 80 years for the gasoline engine.

Don't expect to iron out in an hour all the problems concerning alcohol-powered engines. It might take a day or two.

Appendix

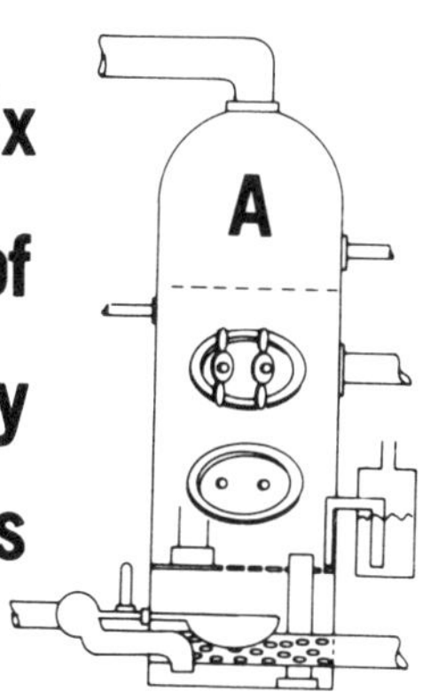

Patent For Production of Acetone And Alcohol By Bacteriological Processes

CHARLES WEIZMANN, OF LONDON, ENGLAND.

PRODUCTION OF ACETONE AND ALCOHOL BY BACTERIOLOGICAL PROCESSES.

1,315,585. Specification of Letters Patent. Patented Sept. 9, 1919.

No Drawing. Application filed December 23, 1916. Serial No. 133,976.

To all whom it may concern:

Be it known that I, Dr. Charles Weizmann, a subject of the King of Great Britain and Ireland, and residing at 67 Addison road, London, W., England, have invented certain new and useful Improvements Relating to the Production of Acetone and Alcohol by Bacteriological Processes, of which the following is a specification.

This invention relates to the production of acetone and alcohols by fermentation processes, and has for its object to obtain large yields by the fermentation of starchy bodies with or without admixture of other carbohydrates in a simple way.

Hitherto the production of acetone and alcohols by the fermentation of starchy bodies has been effected by means of bacteria *inter alia* by bacteria defined as of the type of Fitz. Fermentation of this kind has always been effected under strictly anaerobic conditions in closed vessels.

I have found that certain heat-resisting bacteria, which are identified by the fact that they will convert the greater part of maize or other grain starch into acetone and butyl alcohol, and will also liquefy gelatin, can be used for the purpose of obtaining large yields of acetone and alcohols by the fermentation of solutions or suspensions of natural substances rich in starch or other carbohydrates mixed with such substances under aerobic or anaerobic conditions, *i. e.,* with free access of air as in yeast fermentation, or without.

My invention consists in the fermentation of solutions or suspensions of natural substances rich in starch or of other carbohydrates mixed with such substances by means of the aforesaid heat-resisting bacteria, under aerobic or anaerobic conditions, substantially as hereinafter indicated, with the production of large yields of acetone and alcohols.

The bacteria in question are found in soil and cereals, *e. g.,* maize, rice, flax, etc.

A convenient method of obtaining the bacteria above referred to is as follows:—

I prepare a number, (say 100), of cultures in the usual way by inoculating *e. g.,* hot (say 90²C. to 100² C.) dilute, (say 2%), sterile maize mash with some maize meal, and then allowing it to ferment at about 35² C. to 37² C. for about four to five days.

From these tubes I select those which show the most vigorous fermentation, and have a pronounced smell of butyl alcohol. These selected tubes I now heat up to from 90^2 C. to 100^2 C. for a period of one to two minutes. Many of the bacteria are destroyed, but the desired resistent spores remain. I next inoculate a sterilized maize mash with the culture which has been heated as aforesaid, and so obtain a subculture. I then heat this subculture up to 90^2 C. to 100^2 C. for one to two minutes, and use it to inoculate another sterilized maize mash, and repeat the foregoing subculturing operation a number of times, say 100 to 150 times. In these operations no special precautions need be taken for the exclusion of air.

The bacteria above indicated can then be used in the production of acetone and alcohol under aerobic conditions (which I prefer) by inoculating with the final culture a cooled solution or suspension of the selected substratum, *e. g.,* maize, which has been previously sterilized for three to four hours at a temperature of 130^2 C. to 140^2 C., and a pressure of 2 to 3 atmospheres.

I find that in the case of maize meal a suitable suspension for inoculation can be formed by 100 parts by weight of maize meal, and 1500 parts by weight of water. Fermentation sets in after five to ten hours. The optimum temperature of the fermentation lies about 35^2 C. to 36^2 C.

The fermentation proceeds vigorously for about 36 hours, falling off rapidly after this period, and is completed after a period of about 48 hours. The mash is then distilled, and the products isolated by fractional distillation in the usual way.

I can carry the fermentation under anaerobic conditions in the following manner:—The cooled sterilized maize mash for instance is run into closed tanks in which it can be contained under anaerobic conditions at about 35^2 C. to 36^2 C. It is then inoculated with the hereinbefore described final culture, and allowed to ferment. When the fermentation is completed, the mash is distilled, and the products are isolated as before.

I find that the bacteria above indicated will operate successively on rice, wheat, oats, rye, durra and potatoes as well as maize, and in all these cases without any addition of nutritive materials or stimulants.

The bacteria of the present application, although heat resisting in a sense are not capable of standing the test set by Scheckenbach, in the U.S. Patent 1,118,288, of November 24th, 1914, and are destroyed by any such treatment as are also their spores.

Having now described my invention, what I claim as new and desire to secure by Letters Patent is:—

1. The process of producing acetone and butyl alcohol by the fermentation of liquids containing natural substances rich in starch by means of the herein described bacteria which are capable unaided of converting sterile fermentable grain starch substantially into acetone and butyl alcohol, and also liquefying gelatin.

2. The process of producing acetone and butyl alcohol by the fermentation of liquids containing natural substances rich in starch by means of the herein described bacteria which are capable unaided of converting sterile fermentable grain starch substantially into acetone and butyl alcohol, and also liquefying gelatin, the said fermentation being conducted aerobically as in yeast fermentations.

3. The process of producing acetone and butyl alcohol by the inoculation of a cereal composition with the herein described bacteria which are capable unaided of converting sterile fermentable cereals substantially into acetone and butyl alcohol.

In testimony whereof I have signed my name to this specification.

CHARLES WEIZMANN.

1,315,585

Appendix B
Congressional Testimony

NONDEPARTMENTAL WITNESSES

Agriculture Public Witness Panel

STATEMENT OF PROFESSOR GEORGE T. TSAO, PROFESSOR OF CHEMICAL ENGINEERING AND FOOD AND AGRICULTURAL ENGINEERING, DIRECTOR OF LABORATORY OF RENEWABLE RESOURCES ENGINEERING, PURDUE UNIVERSITY, WEST LAFAYETTE, IND.

ACCOMPANIED BY:

DONALD N. DUVICH, DIRECTOR, DEPARTMENT OF CORN BREEDING, PIONEER HYBRID INTERNATIONAL, JOHNSTON, IOWA

WILLIAM C. BURROWS, SENIOR STAFF SCIENTIST, JOHN DEERE & CO., MOLINE, ILL.

Introduction Of Witnesses

Senator BAYH. Let's return, if you will, please, to the original schedule. I hope you understand why it was important for us to have the Department of Energy in the time frame that we have just utilized. Let's go to the Agriculture Public Witness Panel now, please; Professor George Tsao, professor of chemical engineering and food and agricultural engineering, director of laboratory of renewable resources engineering, Purdue; Dr. Don Duvich, director of the Department of Corn Breeding, Pioneer Hybrid International, Johnston, Iowa; and Dr. William C. Burrows, senior staff scientist, John Deere & Co., Moline, Ill.

I know Dr. Tsao. Dr. Burrows, Dr. Duvich, shall we go from left to right and start off?

Statement of George T. Tsao

Mr. TSAO. Thank you, Senator Bayh. Indeed I feel very honored to have the opportunity to come here to speak about my research at Purdue University. I have prepared a document which I believe has been distributed.

I do have some additional copies in my briefcase.

Prepared Statement

Senator BAYH. We will see that a copy is put in the record. If you could pull that microphone a little closer, it would be helpful.

(The statement follows:)

Prepared Statement of George T. Tsao

Professor of Chemical Engineering and Food and Agricultural Engineering

and

Director of Laboratory of Renewable Resources Engineering

Purdue University

In order to provide a brief technical background, two common chemical terms are first introduced. They are "polymer" and "monomer" (see Fig. B-1). A polymer molecule is made of many monomeric units chemically connected together. The words "poly-ethylene" and "ethylene" illustrate how, in chemistry, polymers are named with respect to their specific monomers. Starch (see Fig. B-2) is a natural polymer that exists in nearly all agricultural grains including corn, wheat and others. The monomeric unit of starch is called anhydro-glucose. Glucose is also known as dextrose and is even better known as grape sugar because glucose is the sugar in grape juice. In wine making, it is glucose that is fermented by yeast into alcohol. A molecule of starch which may also be called poly-glucose is illustrated by the diagram in Fig. B-3 where each letter "G" stands for a glucose unit. One might say that this starch resembles a tree with many branches. The connecting bonds between adjacent glucose units in a starch molecule in this particular state can be considered quite exposed and thus can be easily broken to produce many glucose monomers. In making grain alcohol from corn, starch is first converted to glucose which is then fermented by yeast to produce alcohol.

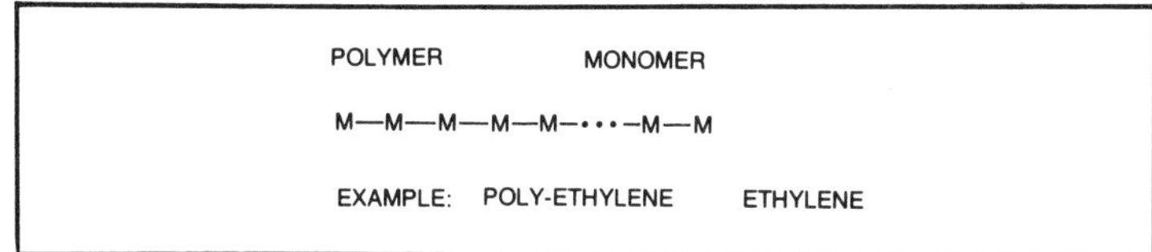

Fig. B-1. Chemical terms.

Table B-1 gives approximate conversion ratios between the tons of alcohol one can produce from the number of bushels of corn consumed in grain alcohol production. At the bottom of this figure, the approximate tonnage of petroleum crude oil consumption of this country is also listed for comparison. These numbers should indicate that grain alcohol should help to reduce our dependence on foreign petroleum if alcohol is used as a liquid fuel in what is called "gasohol," but there is really not enough corn to make a very significant quantity of alcohol as a liquid fuel because the total annual corn harvest of this country is about 6.3 billion bushels weighing about 180 million tons. Adding all other major agricultural crops including wheat, sorghum, etc., the total annual production is about 355 million tons shown in the Table 3-2. This table also gives the figure of some one billion tons of cellulosic wastes available annually in this country, which is actually larger in tonnage than that of petroleum crude oil. In this one billion ton figure, included are cornstalks as opposed to corn grain. It includes bagasse and pulp from sugar mills. It also includes sawmill wastes and small limbs and tree branches from forests. It also includes industrial wastes and big city trash. Some of these materials are readily available and some of them need to be collected obviously at a cost. The common characteristics of all these materials is the fact that they all contain the component called cellulose. Cellulose, like starch, is also a polymer, and its corresponding monomer is also the anhydroglucose unit. Cellulose and starch can both be called "polyglucoses" (Fig. B-4). Unlike starch, however, cellulose is generally difficult to be converted into glucose. Cellulosic materials are of strong and orderly structures that protect cellulose

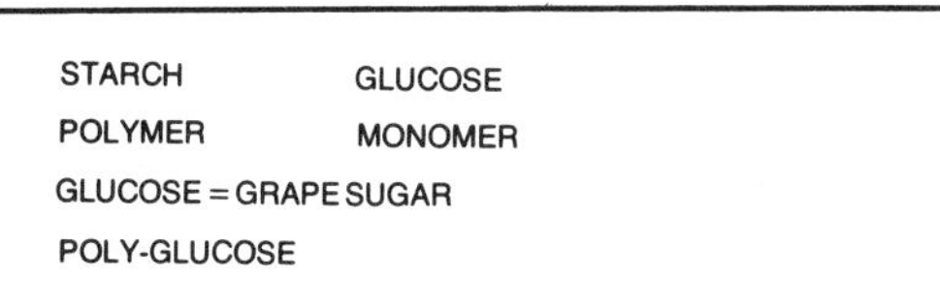

Fig. B-2. Chemical terms.

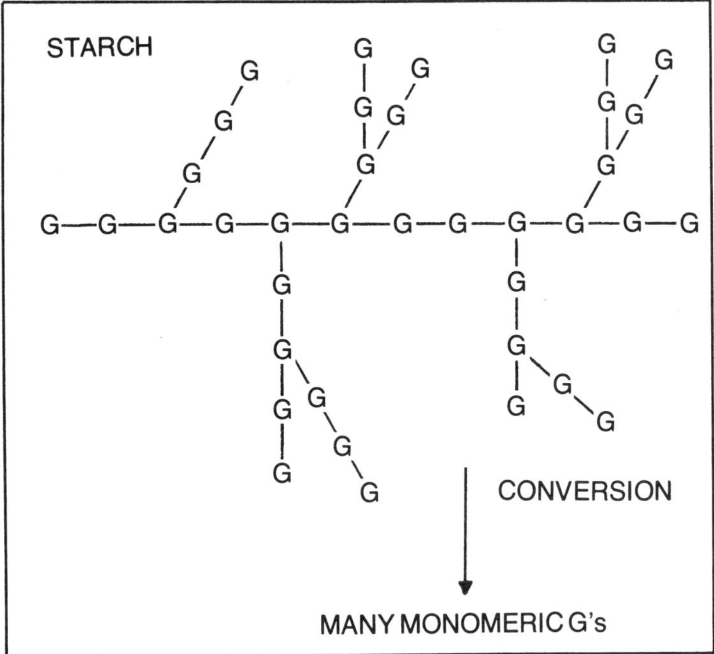

Fig. B-3. A molecule of starch.

from being broken down into glucose. The following several figures are to illustrate these structures. Shown in Fig. B-5, is a model of folded ribbon for a molecule of cellulose polymer. It does not have branches like the one shown previously for starch. A folded ribbon forms a "plate," and as shown in Fig. B-6, several plates on top of one another form an elementary structure unit, each of which has a more or less square cross section. In Fig. B-7, each of the small squares represents one elementary structure unit. In this figure, 16 small squares are shown to form a higher level of structure unit and 64 squares in total to form a yet still higher level of structure unit which is surrounded by a material

Table B-1. Alcohol From Corn.

500 MILLION BUSHELS	4.55 MILLION TONS ALCOHOL
1000 MILLION BUSHELS	9.1 MILLION TONS ALCOHOL
1500 MILLION BUSHELS	13.65 MILLION TONS ALCOHOL
2000 MILLION BUSHELS	18.2 MILLION TONS ALCOHOL
6000 MILLION BUSHELS	54.6 MILLION TONS ALCOHOL
PETROLEUM CRUDE OIL = 820 MILLION TONS PER YEAR	

STARCH POLY-GLUCOSE CELLULOSE

Fig. B-4. Polyglucoses.

called lignin represented by the small crosses. Shown in Fig. B-8, are plant cells that together form a piece of a woody or cellulosic material. The inner hexagonal shaped space on the left half of this figure is full of plasma when the cells are young. When the cells become old, liquid dries up and this inner space becomes hollow. Practically all cellulose of a plant material is contained in the space between the two, inner and outer hexagons. This is the cell wall. On the right side of this figure, an enlarged diagram is showing that the cell wall can be further separated into a primary cell wall p and three distinguishable layers of secondary cell wall S_1, S_2, and S_3. Cellulose exists in these cell walls. In between the cells and surrounding the cell walls is a region called middle lamella which is again made of mostly lignin. Thus, lignin surrounds and thus again protects cellulose at this high anatomical structural level. From the few diagrams in these figures, one should be left with the impression that (1) cellulose has orderly structures and (2) lignin protects cellulose by surrounding it. Indeed, as summarized in Fig. B-9, these are the two major obstacles to converting cellulose in order to obtain glucose.

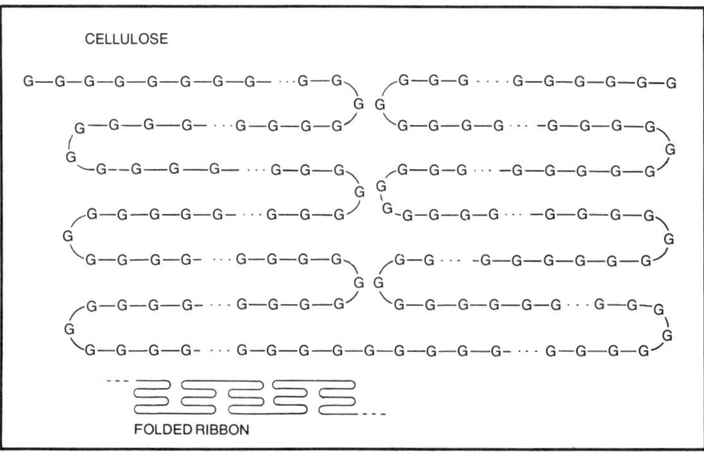

Fig. B-5. A molecule of cellulose polymer.

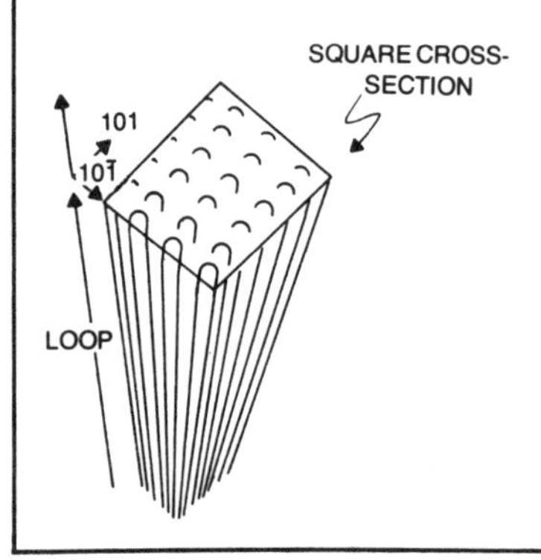

Fig. B-6. Perspective view of a cellulose crystallite according to the Chang model.

The idea of obtaining glucose from cellulose is not new and technologists have tried for over 50 years. Recently, we at Purdue University have worked out a new process which gives 100% yield of glucose from available cellulose at a low cost. As diagrammed in Fig. B-10, a solvent A is first used to remove the third major component of cellulosic materials, besides cellulose and lignin, called hemicellulose. The solid residue containing cellulose and lignin is then treated with another solvent B which dissolves cellulose but not lignin. This approach which is simple in concept will obviously remove the two major obstacles to cellulose conversion as described by the structure diagrams in the last several figures. As shown in Fig. B-11, the cellulose dissolved in solvent B can be made to re-precipitate by mixing with water. Solvent B is to be recovered for recycle and the re-precipitated cellulose which is by now no longer protected by lignin and has no

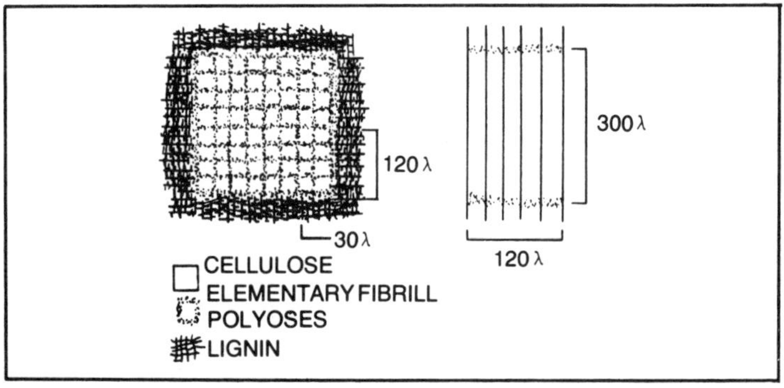

Fig. B-7. Model of the ultrastructural organization of the cell wall components in wood; cross section on the left, longitudinal section of the right.

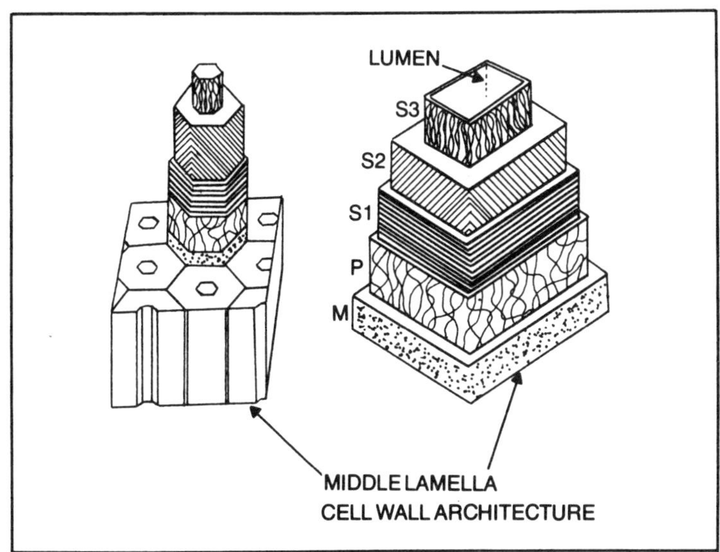

Fig. B-8. Plant cells.

strong and orderly structure can be easily converted to give 100% yield of glucose.

In summary, as shown in Fig. B-12, from both starch in grain and cellulose in crop residues, glucose can be derived. Glucose can be fermented by yeast to produce alcohol and can also be processed to produce many chemicals replacing petroleum. From a ton of grain corn, about three tenths a ton of alcohol has been produced. With the new treatment technique involving solvents, one can potentially produce up to four tenths a ton of liquid alcohol from a ton of dry cornstalks. The work at Purdue University has been supported by the university administration, by previously AEC, then ERDA and now DOE and also by the National Science Foundation. For this generous support, I am very grateful. We at Purdue University are now at a stage of continuing our laboratory research. We are seeking for funds to study our process on a somewhat larger scale. There are several technical tests which

TWO MAJOR OBSTACLES TO CELLULOSE HYDROLYSIS

1. STRONG AND ORDERLY STRUCTURE OF CELLULOSE
2. PROTECTION BY LIGNIN SEAL

Fig. B-9. Obstacles to converting cellulose.

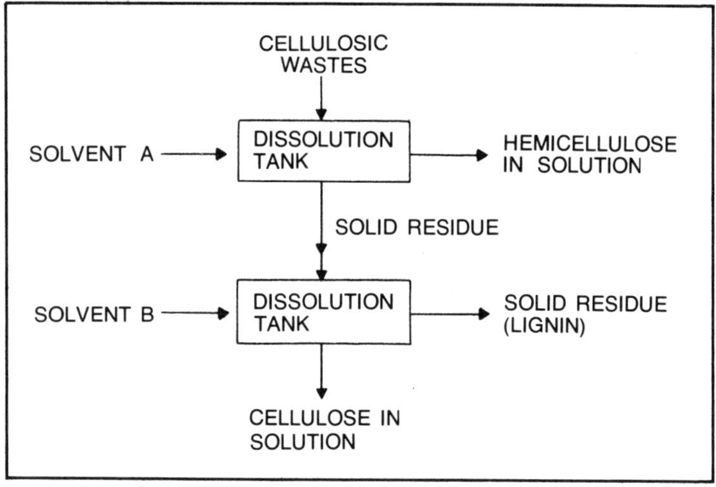

Fig. B-10. Solvent extraction-scheme 1.

cannot be meaningfully done in test tubes and laboratory scale glassware. If we are given a grant of 1.5 million dollars, we can complete the process research on a pilot plant scale in about two years, after which we will have data for detailed economic and energy analysis and for designing production scale factories. We need another 1.0 to 1.5 million dollars to study problems such as the techniques of collecting, compacting and transporting cellulosic materials, re-designing the necessary farm machineries, agronomy and soil research, and the potential long term and short term economical impact.

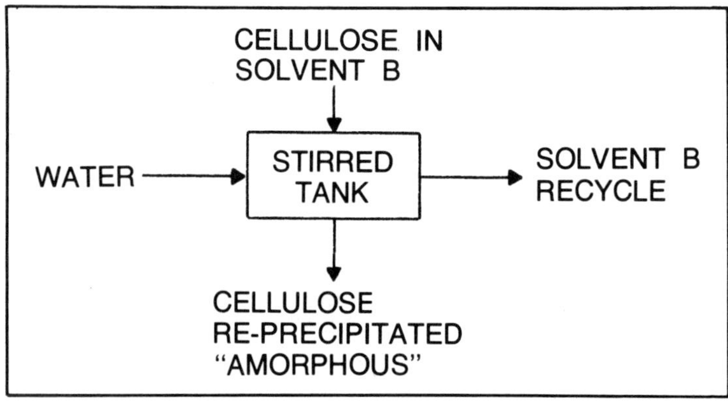

Fig. B-11. Dissolved cellulose can be made to re-precipitate by mixing with water.

Based upon our laboratory results, a crude cost analysis was made for the process of conversion of cellulosic wastes to glucose and fermentable sugars. The results are based upon a cost of $30 per ton of cornstalks purchased from farmers. With our treatment techniques, we believe that we can produce crude grade glucose at about 1.8 cents per pound which will mean that alcohol can be produced at about 74 cents per gallon.

Some of the cellulosic wastes such as urban trash and animal feedlot wastes actually have a negative values and thus do not cost $30 per ton to collect. Some of the other wastes such as bagasse from sugarcane are already collected at sugar mills. They are currently burnt for heating value. Because of its high moisture content, the heating value is low and is worth only about $7.50 per ton for direct burning. Using a more optimistic approach by assuming no cost instead of $30 per ton for the cellulosic raw materials and a reasonable return on selling the yeast byproduct from alcohol stillage, one can then come up with an even more attractive cost figure for alcohol production at about 40 cents per gallon.

Alcohol is only one of many products that can be made from glucose. Table B-3 is a partial list of additional products that can be derived from glucose. Some of these are direct consumer products and some of these are industrial intermediates which in turn can be processed into even a larger number of products. From the other two major components of cellulosic materials, namely hemicellulose and lignin, one can also make a large number of petroleum sparing products. These are illustrated in Table B-4 and Table B-5. Indeed, if we have to, complete processing industries can be developed based upon the ultimate raw material, cellulosic materials, which is a valuable renewable resource.

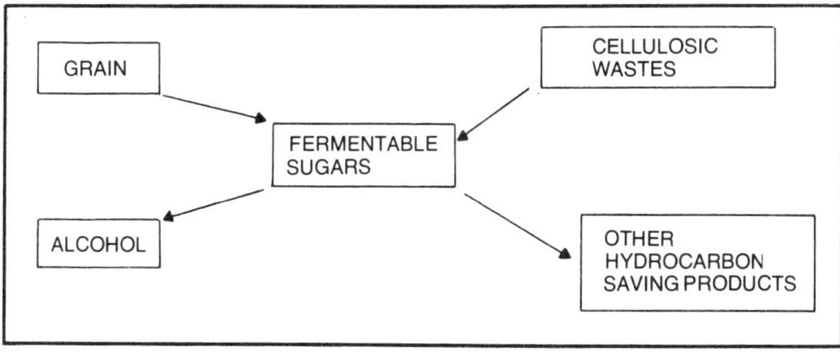

Fig. B-12. Glucose can be derived from starch in grain and from cellulose in crop residue.

Table B-2. Annual Production.

PETROLEUM CRUDE OIL	820 MILLION TONS PER YEAR
GRAINS	355 MILLION TONS PER YEAR
CELLULOSIC WATERS	1010 MILLION TONS PER YEAR

Finally, I would like to share with you a brief discussion I had with an Indiana farmer. With the depressed prices for corn and other farm products in current market, farmers will be happy to obtain $30 per ton for their crop residues such as cornstalks. For instance, our national average of corn harvest is about 80 bushels per acre. At $1.50 per bushel of current price, this yields a gross income of only 120 dollars per acre. For each acre of corn field, there is a yield of two to three tons of cellulosic residues per year. Saving a portion of the residues for soil conditioning, in average, one can remove about one ton of crop residues per acre per year from our farms. This ton of residue will yield an extra gross income of $30 per acre per year which is substantial. See Fig. B-13. The "land set aside" program costs our government large sums of tax money. The acreage can be used for producing cellulosic materials sold at $30 per ton. My farmer friend told me that, for an acre of land, if he plants alfalfa he can harvest about 7 tons of dry cellulosic materials contained per year. At $30 per ton, this can give him a better income than planting corn. The wet, freshly harvested alfalfa, if pressed immediately, can produce a juice which contains high quality protein as a byproduct. The fibrous residue after juice removal is where the cellulose is. Furthermore, alfalfa has the capability of fixing nitrogen from atmosphere. If farmers rotate their crops between alfalfa and corn, they can save on the nitrogen

Table B-3. Products Derivable From Cellulose.

Cellulose

↓

Fibers
Pulp and Paper
Glucose ⟶

Fructose sugar	Acetic Acid
Ethanol	Polysaccharides
Butanediol	Vitamins
Antibiotics	Glycerine
Amino Acids	Acetone
Synthetic Rubber	Lactic Acid
Synthetic fiber	Plastics
Citric Acid	Single Cell Protein

Table B-4. Products Derivable From Hemicellulose.

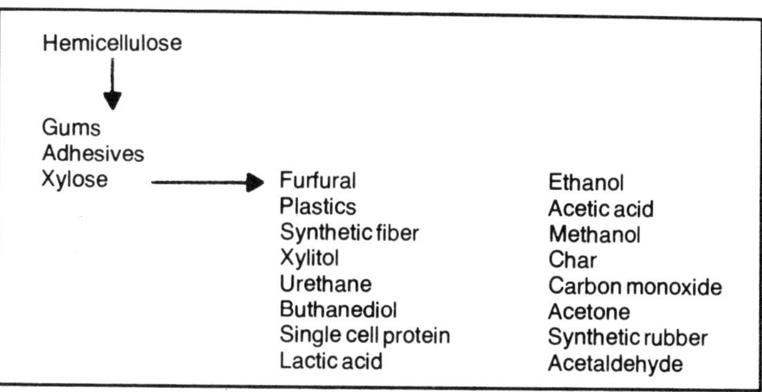

fertilizer. Corn, unlike soybean and alfalfa, has no nitrogen fixative capability and requires heavy application of ammonia as its nitrogen source. In average, one third of the energy consumption in farming of corn is for producing the necessary nitrogen fertilizer, ammonia. The energy consumption for this is equivalent to 26 gallons of gasoline per year per acre of corn. My farmer friend does not have answers to all my questions, but he has certainly pointed out the need of much integrated research for both food and energy production. Our country is blessed with the largest and the richest land mass in the world. It is a shame to all of us, today we will pay our farmers for not farming their land; while we spend tens of billions of dollars every year to buy foreign oil. We should all ask ourselves how long can this country afford to let this go on.

John Deere & Co.
Statement Of William C. Burrows

Mr. BURROWS. Thank you very much, sir.

Table B-5. Products Derivable From Lignin.

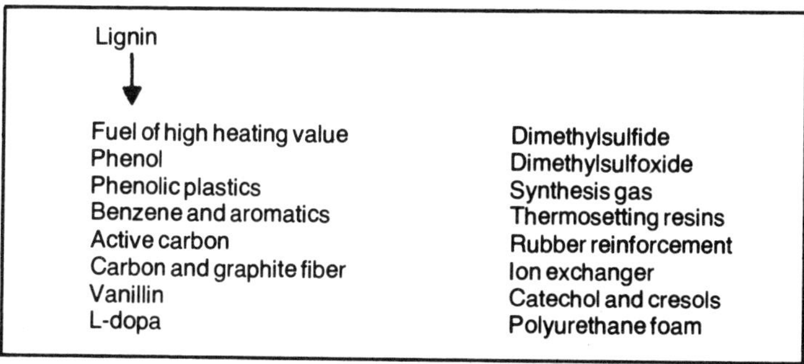

50 TONS CROP RESIDUE ——► 79,000 LBS SUGAR	
CRUDE COST ANALYSIS	
1. 50 TONS RESIDUE AT $30 PER TON	$1500
2. CHEMICALS	215
3. ENERGY	156
TOTAL COST	$ 1871
CREDIT FOR ENERGY FROM BURNING LIGNIN	475
NET ABOUT	$ 1400
OR 1.8 CENTS PER POUND OF FERMENTABLE SUGAR	
74 CENTS PER GALLON ALCOHOL	

Fig. B-13. Cost analysis.

Very interesting comments by Dr. Tsao. We are certainly pleased we have been invited to testify at these important hearings because our studies have convinced us of the necessity for all possible haste in dealing with the energy problems that our Nation faces in the near future.

Of course, we are connected with Agriculture.

Agriculture has an especially critical need for an assured supply of portable fuel. I have made available to the committe a study of ours which gives a perspective on the critical nature of portable fuels in agriculture.

Some may question how we define the near future. From the standpoint of a manufacturer of farm and industrial equipment, I would point out that a great many of the tractors in use today are at least 20 years old. Thus, the tractor that rolls off the assembly line today will be expected to do useful work in 1998. So to us, and to the purchaser of this tractor, 20 years is one tractor lifetime and that is pretty near. And this tractor is designed to run efficiently on today's fuels.

Fuel And Engine Development

Fuel and engine developments will continue to proceed hand in hand as they have in the past. Engine technology and economics depend on the end use of the engine. For example, the gas turbine engine has been highly developed for aircraft use, but it is not an economically or technologically viable option for use in tractors, even after more than 20 years of research and development.

Conversely, diesel engines—incidentally, we are completely diesel engine oriented—which meet the economic, technical and operating conditions required for farm equipment, will probably never be suggested for use in aircraft.

While the development of a new engine may take 20 years or more, the redesign of a current engine takes much less time. For example, some of the improvements in the diesel engine we have seen in the recent past, such as turbocharging, require about 4 to 6 years lead time.

Redesigning an engine line to include higher displacement and higher power models may require up to 8 years lead time. But these are still diesel engines. This addresses, in part, the question of modifying current engines to make use of alcohol fuels in farm equipment.

As I mentioned previously, engine and fuel developments go hand in hand so the question becomes what kind of fuel-engine combinations will be viable? Our studies have shown that, with the likely availability, well past the year 2000, of diesel fuel made from coal, the inherent efficiency of the diesel engine gives it a clear advantage for use in farm and industrial equipment in the future.

Prepared Statement

At this time, Mr. Chairman, I would like to advise you that I have furnished to the committee staff copies of two other studies referred to here as well as copies of the oral testimony. I request permission to make these documents part of the record of our views. They contain many details too lengthy for the oral presentation.

Senator BAYH. We will incorporate those with your statements.

Mr. BURROWS. Thank you, sir.

[The information follows:]

Prepared Statement of W. C. Burrows

We are pleased to have been invited to testify at these important hearings. Our studies have convinced us of the necessity for all possible haste in dealing with the energy problems our nation faces in the near future. Agriculture has an especially critical need for an assured supply of portable fuel. I have made available to the Committee a study of ours which gives a perspective on the critical nature of portable fuels in agriculture. Some may question how we define near. From the standpoint of a manufacturer of farm and industrial equipment, I would point out that a great many of the tractors in use today are at least 20 years old. Thus, the tractor that rolls off the assembly line today will be expected to do useful work in 1998. So to us, and to the purchaser of this tractor, 20 years is one tractor lifetime. And this tractor is designed to run efficiently on today's fuels.

Fuel and engine developments will continue to proceed hand-in-hand as they have in the past. Engine technology and economics depend on the end use of the engine. For example, the gas turbine engine has been highly developed for aircraft use, but it is not an economically or technologically viable option for use in tractors, even after more than 20 years of research and development. Conversely, diesel engines, which meet the economic, technical and operating conditions required for farm equipment, will probably never be suggested for use in aircraft.

While the development of a new engine may take 20 years or more, the re-design of a current engine takes much less time. For example some of the improvements in the diesel engine we have seen in the recent past, such as turbocharging, require about four to six years lead time. Redesigning an engine line to include higher displacement and higher power models may require up to eight years lead time. But these are still diesel engines. This addresses, in part, the question of modifying current engines to make use of alcohol fuels in farm equipment.

As I mentioned previously, engine and fuel developments go hand-in-hand so the question becomes what kind of fuel-engine combinations will be viable? Our studies have shown that, with the likely availability, well past the year 2000, of diesel fuel made from coal, the inherent efficiency of the diesel engine gives it a clear advantage for use in farm and industrial equipment in the future.

At this time, Mr. Chairman, I should like to advise you that I have furnished to the Committee Staff copies of two other studies referred to here as well as copies of the oral testimony. I request permission to make these documents part of the record of our views. They contain many details too lengthy for the oral presentation.

Our assumption that diesel fuel will be available in the future should be examined more closely. And it is at this point that we must consider alternate fuels such as alcohol. Alcohol has a high octane rating, similar to gasoline. This means that it has a low cetane number, which is the basis for rating fuels used in compression ignition engines such as the diesel engines. But the characteristics of alcohol do not make it a good candidate for diesel engine fuel. Pure alcohol will not ignite as does diesel fuel. If the engine is provided with some ignition system, alcohol will burn in such an engine, however. Using alcohol as a fuel with current diesel engine technology leads to less efficient operation of the engine.

Even though we do not consider currently proposed alternate fuels to be directly applicable to diesel engines without modification does not mean that we are ignoring these possibilities. We have examined the use of alcohol as a fuel in spark ignition engines. We are also investigating what it means to use the gaseous products of pyrolysis, as well as alcohol, as the fuel for diesel engines.

This specifically addresses the question of the modifications required in diesel engines to use alternate fuels. At this point we know that modifications will be necessary to use alcohol as a fuel in farm equipment and that there are several major technical problems to be solved. For example, the method of delivering alcohol fuel to the engine is one. Diesel fuel is delivered to the engine by an injection pump, which also provides the necessary timing.

The use of alcohol as the primary portable fuel for diesel engines thus demands two things. First, the alcohol must be ingested into the engine through an alcohol-tolerant supplementary injection system or it must be added to the inlet air, not unlike a carburetted gasoline fueled engine. Ignition timing can be provided by using the present diesel fuel injection system for pilot ignition injection with a small percentage of diesel fuel or a totally separate ignition system can be provided. In some cases ignition improving additives must be added to the alcohol to insure reliable ignition. To bring the engine back to equivalent power density of current diesel engines may also require different combustion chamber configuration.

All the above require a modified engine design. We believe it is possible to design a modification kit which will field convert diesel engines to alcohol use. At the present time, however, we do not have a modification kit available and do not know how effective field modification of engines would have to be to make the conversion economically attractive. This modification can become attractive only when alcohol becomes available as a competitive portable fuel. We do believe that conversion is possible and considering all the alternatives might make an economically practical package.

Agricultural residues, surplus agricultural production, forest residues and the plant material growing on marginal lands are the largest portion of what is collectively referred to as biomass. It represents a sizable, and renewable energy resource, equivalent to a part of total U.S. petroleum use. We consider that biomass will be used as an energy source in the future. Machinery is currently in production and available to processors to handle the growing and harvesting of biomass for energy. We think this applies both to the agricultural as well as the forestry area.

We recognize, of course, that coal is a good source of alcohol. However, coal can also be a good source of synthetic crude oil from which both gasoline and diesel fuel can be derived. Considering the ease of substituting alcohol for gasoline and the difficulties involved in replacing diesel fuel with alcohol, we feel that if coal is to be converted to portable fuel, the most appropriate fuel is diesel and that alcohol for automotive use be derived from the renewable resources.

One of the basic questions this hearing deals with is what Federal policy would be helpful in facilitating the earliest possible use of alternate fuels as a substitute for petroleum-based fuels. Earlier I mentioned that fuel and engine developments are made together. An engine manufacturer would be foolish to manufacture an engine for which an assured supply of the most appropriate fuel is not available at a reasonable cost to the user. The same applies to the fuel producer. Both of these industries currently have the technical and economic resources to develop the most efficient-fuel combination for the future.

What is really needed to encourage research into alternate fuels is two things: First, a Federal energy policy which will let free market forces govern the price and availability of portable fuels. No major industrial enterprise will risk the enormous amounts of capital necessary to develop alcohol as a major alternate portable fuel when return on this investment cannot be assured by the economics of the free market. The second thing that is important, and in a way is a corollary of the first requirement, is for the Federal Government to define the technical legislative framework within which future self-powered vehicles must operate. For example, what will be the limits on noise, exhaust emissions, raw fuel toxicity, and operator, bystander and producer safety? Some concern must also be expressed should environmental protection legislation focus on the collection of biomass—not unlike the current concern with strip-mining of coal. This technical legislative framework must be clearly defined and understood if major investments are to be encouraged in the private sector.

The research capabilities of the Federal Government and the State experiment stations in the areas of agriculture and forestry are unique for answering many of the questions involved with long term use of renewable resources for energy. For example, the effects of mining the soil by continual removal of biomass materials. A good start has been made by USDA's residue research team headed by Dr. William E. Larson at St. Paul, Minnesota. We think this program should receive added emphasis. Another question involves the consequences of alternative uses of land for energy, food, feed and fiber on food supply, food cost, balance of payments, etc. These are but two areas - many more may come to mind. Private enterprise is not equipped to conduct research into this type of question. Rather, they could be a source of information required to finally develop the best overall system that considers all aspects of national needs, including energy.

A most confusing aspect of using either coal or biomass-derived fuels for engines is the multiplicity of technologically feasible proposals now on the table. Examples include methanol, methane, hydrogen, producer gas, ethanol, and direct combustion for fueling engines, heating buildings, manufacturing process energy, electricity generation, chemical feedstocks, etc. Each proponent of a process or use seems to feel most or all of the available resources would be utilized for that project. Obviously, priorities will be set in the future. However, very little investment will or should be made in this uncertain climate. A clear set of priorities would be helpful. This is an area where the Federal Government should play a role with the assistance and input of the private sector. In the final analysis, it must be everyone, both in the public and private sectors working together to solve our dilemma.

Future Availability Of Diesel Fuels

Mr. BURROWS. Our assumption that diesel fuel will be available in the future should be examined more closely. And it is at this point that we must consider alternate fuels such as alcohol. Alcohol has a high octane rating, similar to gasoline. This means that it has a low cetane number, which is the basis for rating fuels used in compression ignition engines such as the diesel engines.

But the characteristics of alcohol do not make it a good candidate for diesel engine fuel. Pure alcohol will not ignite as does diesel fuel in a compression ignition engine.

If the engine is provided with some ignition system, alcohol will burn in such an engine, however. Using alcohol as a fuel with current diesel engine technology leads to less efficient operation of the engine.

Even though we do not consider proposed alternate fuels to be directly applicable to diesel engines without modification does not mean that we are ignoring these possibilities. We have examined the use of alcohol as a fuel in spark ignition engines. We are also investigating what it means to use the gaseous products of pyrolysis, as well as alcohol, as the fuel for diesel engines.

Problems Associated With Alcohol Use In Farm Equipment

This specifically addressed the question of the modifications required in diesel engines to use alternate fuels. At this point, we know that modifications will be necessary to use alcohol as a fuel in farm equipment and that there are several major technical problems to be solved.

For example, the method of delivering alcohol fuel to the engine is one. Diesel fuel is delivered to the engine by an injection pump, which also provides the necessary timing.

That means you don't have a distributor on a diesel engine.

The use of alcohol as the primary portable fuel for diesel engines thus demands two things. First, the alcohol must be ingested into the engine through an alcohol-tolerant supplementary injection system or it must be added to the inlet air, not unlike a carburetted gasoline fueled engine. Ignition timing can be provided by using the present diesel fuel injection system for pilot ignition injection with a small percentage of diesel fuel or a totally separate ignition system can be provided as is provided in some engines that have been researched at other locations.

In some cases, ignition-improving additives must be added to the alcohol to insure reliable ignition. To bring the engine back to equivalent power density of current diesel engines may also require different combustion chamber configuration.

Engine Modification Kit Development

We have been talking about things that require modified engine design. We believe it is possible to design a modification kit which will field convert diesel engines to alcohol use. At the present time, however, we do not have a modification kit available and do not know how effective field modification of engines would have to be to make the conversion economically attractive.

This modification can become attractive only when alcohol becomes available as a competitive portable fuel. We do believe that conversion is possible and considering all the alternatives, might make an economically practical package. I would like to switch now to more basic agricultural things.

Energy Development From Biomass

Agricultural residues, surplus agricultural production, forest residues and the plant material growing on marginal lands are the largest portion of what is collectively referred to as biomass. It represents a sizable and renewable energy resource, equivalent to a part of total U.S. petroleum use.

We consider that biomass will be used as an energy source in the future. Machinery is currently in production and available to processors to handle the growing and harvesting of biomass for energy.

Maybe I should emphasize that a little more because it is easy to pass over this in the testimony. We think the technology, the machinery for using this material is here now. It can be bought in the marketplace. We also think that this applies both to the agricultural and the forestry area.

Coal A Source Of Alcohol

We recognize, of course, that coal is a good source of alcohol. However, coal can also be a good source of synthetic crude oil from which both gasoline and diesel fuel can be derived. Considering the ease of substituting alcohol for gasoline and the difficulties involved in replacing diesel fuel with alcohol, we feel that if coal is to be converted to portable fuel, the most appropriate fuel is diesel and that alcohol for automotive use be derived from the renewable resources.

Federal Policy In Alternate Fuel Development

One of the basic questions this hearing deals with is what Federal policy would be helpful in facilitating the earliest possible use of alternate fuels as a substitute for petroleum-based fuels.

Earlier, I mentioned that fuel and engine developments are made together. An engine manufacturer would be foolish to manufacture an engine for which an assured supply of the most appropriate fuel is not available at a reasonable cost to the user. The same applies to the fuel producer.

Both of these industries currently have the technical and economic resources to develop the most efficient-fuel engine combination for the future.

What is really needed to encourage research into alternate fuels is two things: First, a Federal energy policy which will let free market forces govern the price and availability of portable fuels.

No major industrial enterprise will risk the enormous amounts of capital necessary to develop alcohol as a major alternate portable fuel when return on this investment cannot be assured by the economics of the free market.

The second thing that is important, and in a way is a corollary of the first requirement is for the Federal Government to define the technical legislative framework within which future self-powered vehicles must operate. For example, what will be the limits and noise, exhaust emissions, raw fuel toxicity, and operator, bystander and product safety?

Some concern must also be expressed should environmental protection legislation focus on the collection of biomass—not unlike the current concern with strip-mining of coal. This technical legislative framework must be clearly defined and understood if major investments are to be encouraged in the private sector.

Research Capabilities of Federal And State Governments

The research capabilities of the Federal Government and the State experiment stations in the areas of agriculture and forestry are unique for answering many of the questions involved with long-term use of renewable resources for energy. For example, the effects of mining the soil by continual removal of biomass materials. A good start has been made by USDA's residue research team headed by Dr. William E. Larson at St. Paul, Minn.

We think this program should receive added emphasis. Another question involves the consequences of alternative uses of land for energy, food, feed, and fiber on food supply, food cost, balance of payments and so forth.

There are but two areas—many more may come to mind. Private enterprise is not equipped to conduct research into this type of question. Rather, they could be a source of information required to finally develop the best overall system that considers all aspects of national needs, including energy.

Proposals For Coal Or Biomass Uses For Fuel

A most confusing aspect of using either coal or biomass-derived fuels for engines is the multiplicity of technologically feasible proposals now on the table.

Examples include methanol, methane, hydrogen, producer gas, ethanol and direct combustion for fueling engines, heating buildings, manufacturing process energy, electricity generation, chemical feedstocks, and so forth.

Establishment Of Priorities

Each proponent of a process or use seems to feel most or all of the available resources would be utilized for that project. Obviously, priorities will be set in the future. However, very little investment will or should be made in this uncertain climate.

A clear set of priorities would be helpful. This is an area when the Federal Government should play a role with the assistance and input of the private sector.

In the final analysis, it must be everyone, both in the public and private sectors working together to solve our dilemma.

Thank you, sir. That is the end of my prepared statement.

Senator BAYH. Thank you, Mr. Burrows. That has been very helpful.

Before asking you any questions, Mr. Duvich, why don't you let us have your thoughts? Then we will question all three of you gentlemen, if you don't mind.

Development Of Alternate Cellulosic Crops

Senator BAYH. Dr. Duvich, in Professor Tsao's testimony, he used the example of the conversation he had with the farmer. I think the figure he used was 7 tons of alfalfa and he talked about the alcohol equivalent, and the protein extract which would be a valuable by-product.

Is it fair to assume that if the Nation determines that it is in the national interest to develop an alternative source of energy that is not petroleum-based, to take advantage of the productive capacity of the American farmer that, whether it is agriculture research or university research, that it is entirely possible that a product can be grown that has a significantly higher cellulosic content than even alfalfa.

Mr. DUVICH. It may well be. You are thinking of some crop other than the ones we currently grow in the Midwest or are you thinking, looking through the Midwestern crops, or any crop in the United States for that matter, that is now grown on the farms?

Senator BAYH. I am looking at the entire Nation. I don't know what the answer is. We have crops now being produced that have cellulosic value. But has there really been any effort made to try to see how much cellulose we can produce per acre, how much alcohol we can produce per acre?

In your testimony, you note that what we have been doing on the farm is to try to produce more grain and less stock.

Mr. DUVICH. I see your question and it is an excellent one because our agricultural production and the breeding work, the cultivation has been pointed towards almost anything except cellulose. I think it could well be by either looking into the crops that we now grow or perhaps by looking toward new crops that we have not yet considered we would find things which could produce much more cellulose per acre than is currently available.

Plant Breeding

Senator BAYH. One of our witnesses—I think one of the USDA folks, spoke of some research on energy sources which are designed specifically with cellulose content in mind.

Tell me this—I should know from my school days but I have to confess that my memory is not too good—in the structure of cellulose, as it exists in certain plant fibers, is it possible to breed—I guess I get back to you, Mr. Duvich, and the plant breeding—is it possible to breed a plant that will have the same size cornstalk, or the same ratio of stock to ear, that we have now but with the cellulose content of the stock higher?

I know we have done things to increase protein in grains. Can we do the same thing to increase the protein value, protein output of a given amount of stalk?

Mr. DUVICH. To increase the cellulose amount of a given amount of stalk is the question?

Senator BAYH. Yes.

Mr. DUVICH. Probably yes. Certainly the internal structure of the corn stalk varies a great deal from one hybrid to another, from one gene type to another.

A lot of this certainly must be reflected in the cellulose content of it. Perhaps Dr. Tsao could say more about it?

Mr. TSAO. This is outside of my area of expertise, but you have talked to my colleagues at the university who have worked in the agronomy department. He is doing research to look for new kinds of corn which has a lower content of lignin.

As I recall from the diagram I showed earlier, lignin is the one that makes cellulose hard to break down to look at. So now, research is going on to look for things which contains less lignin, comparatively speaking, than for the higher cellulose. That type of thing is indeed possible.

Stronger Genetic Breeding Capabilities

Senator BAYH. I know this is a long process. It takes awhile. We have done remarkable things, breeding genetic capabilities to resist disease, and to resist certain kinds of weather, to stand longer and to avoid problems so the darned things will pop out; to breed the characteristics of certain varieties, the kind that will stay there all winter.

Mr. TSAO. I might also add that the income figure I quoted earlier, that there is an amount of solid residue which already is available, already is produced as a byproduct for some other purpose.

I believe the Department of Energy, the program that Dr. Roscoe Ward, who was here awhile ago, has been looking into various kinds of other possibilities. They speak in terms of the energy farming, farming specifically for the purpose of producing energy.

That means for things like land, which is not good enough for food production, but it could be very well useful for producing biomass for energy. Also, for instance, this is taking place in fact in our neighboring Brazil. Because of the price of sugar, the Brazilians now are talking about taking a whole sugarcane for the purpose of producing alcohol rather than producing sugar.

It turns out sugarcane probably, as of now, by far is the best producer of biomass. It produces something like 20 dry tons per year per acre as compared with, say, 7 tons for alfalfa. Of course, sugarcane has a limitation on the weather conditions. It is one of the things you have to consider in certain areas.

Also, I believe work has been going on, on producing green algae for the purpose of biomass production and this type of green algae can grow again very fast. It utilizes, for instance, the sea body, the water body near the coastal areas, the areas that you can't use for food crops but, indeed, you can use those areas for producing biomass for fuel.

Priority Of Tillable Acreage

Senator BAYH. It is sort of a delicate balance here. We only have so many acres that can be tillable. Certain acres are tillable for certain kinds of crops and not for others. Others that are tillable are good for producing certain kinds of crops and not for others.

If our long-range goal is to have a significant amount of our energy developed from a renewable base here, we have to set some priorities on the use of our acreage, so that those acres that are exceptionally good for growing corn or soybeans, be utilized for that; others suitable for wheat but not for corn and soybeans should be used for that, and so on. I believe I read somewhere that a fast growing poplar, may grow more cellulose than sugarcane. We just don't know the answer yet.

Let me ask you, Mr. Burrows, could you tell us a little bit more about the technology that is now available? Let's look at products that are now wasted, for all intents and purposes, that have no commercial value. Although they do have an organic value to the farm that we cannot ignore, all of it is not needed. What technology is available now, in the way of farm implements to make it possible to handle that large volume, low-weight commodity in such a way that the very cost of handling it doesn't make the process prohibitive?

Mr. BURROWS. Sir, certainly what you say is true. We can't use all that is produced and that is why I mention that one particular USDA study, incidentally, because I think we have got a chance for the first time to get a good estimate of exactly how much we can remove.

Senator BAYH. Would you excuse me for interrupting? I want to say I don't want to be discourteous, but I am going to have to sit here until 2:30 or 3 o'clock and then move to another forum. If I may, I would like to sort of nibble here.

Mr. BURROWS. I saw it delivered. I was smiling.

Senator BAYH. It looks better from where you are sitting than it does from where I am sitting. I don't want to be discourteous, but I hope you will understand the necessity.

Mr. BURROWS. Yes, I understand completely.

As I said, for the first time I think we can determine about how much we can remove and I think this is critical from the standpoint that we must know how much we can get from where, in order to have some idea of how far we have to transport this material, and thus how large a conversion facility can be and how much area needs to be covered to supply this plant.

Senator BAYH. Let me be a little more specific here. You are in the machinery business. You do a pretty good job of it. You have come a long way since those old 2-cylinder jobs that put me to sleep while I was running them.

Mr. BURROWS. They happened to be very fuel efficient.

Mechanical Harvesting Of Cornstalks

Senator BAYH. And very good. Right. What is available now, what can be done, and in how long a period of time, to do a more sophisticated and efficient job of harvesting cornstalks as well as corn.

Mr. BURROWS. There is machinery presently available on the market. I can speak for our own company, other companies have similar machinery in some cases. We have a header for a machine. It goes on our Forage Harvester currently, but it can be adapted.

That is called a stalker. That is our trademark name for it. This machine will harvest four rows of cornstalks at a time, leave a small amount there. You can set it so you can leave a little more, so you can harvest about as much as you wish.

The material can then be thrown back into a forage wagon, some kind of a wagon. Alternatively, it can be put into what we call a stack wagon. There are other names, trademark names for this kind of a machine.

It is a wagon that makes large stacks of the material, which ordinarily at the present time is used for hay, although some farmers use it for stalk, making stalk, putting up stalking, for example, for cattle feeding. These increase the density of the material.

It is a large wagon that the material is blown into and the roof of the wagon is hydraulically operated, comes down and squeezes the material down and you pack it in there. That is probably the least dense type of package, not considering just the plain wagon loaded material.

Baling Machines

Then the regular balers will work. The plain hay balers will work for baling stalks and any other material that you might get as a residue.

The large, round bale is a thing that is becoming more and more common on the American farm scene. I am sure you have seen them as you have driven down the highway in Indiana. This gives you just a little bit higher density than the stack wagon does.

There are machines that will handle these. They weigh up to about 1,700 to 1,800 pounds per bale.

There is another machine that is newly on the market that we don't happen to make. Easton makes this machine. That is a large square baler. It will put up even denser bales, more like the very dense bales we used to see and you still see out in the West, out in Nebraska.

You see 80-pound bales, not 40-pound ones that we are used to throwing around in the Midwest.

Further than that, there is a machine that John Deere makes, called a cuber, which gives you a density that is higher than any baled material, but slightly less than the pellet, sort of like rabbit pellets or chicken feed pellets.

The cuber will increase the density. The material that is made by this, 2 by 2 inches by about 2½ inches long, still called a cube. It can be handled in much the same way that grain will be handled. It can be put in the same kind—it can be augered. It is better to use a chain elevator, or something like that.

Then, of course, the pelleting machines that have been common for years and years and are currently being used in some forms of biomass energy production, particularly the Woodex process out in Oregon that we have heard about.

Senator BAYH. Let me ask you to hypothesize here. You are talking about 1978 technology.

Mr. BURROWS. That is what is available. I don't see that there is any real necessity for much of anything else, but I could be wrong.

Senator BAYH. Let me think a minute. All right? I see some of my farm friends driving the kind of combines that are combination combines and cornpickers, with some of the conveniences that have now become necessities.

When I was pulling my old two row or even my mounted two row out, suffering near frostbite, I certainly did not foresee the necessity of what people now call necessities. You fellows in the farm machinery area have done a fantastic job.

Let us assume that we are looking at the long-range goal here. Let's say we are going to accept commodity χ or commodities A, B, C, D as part of a renewable energy resource program.

Part of that is cornstalk. I would assume that if there is a market for conrstalks, that you are going to come up with some sort of contraption that is going to harvest cornstalks in the most efficient manner to get them to market and will permit one harvesting operation, eliminating the need to go back over the field again.

Is there any way of assessing what we are talking about as far as costs? What is that going to mean?

Per Acre Value Of Cornstalks

Dr. Tsao, you talk about $30 an acre. Does that include the cost of harvesting? And have you given any thought—I think maybe this is one area in which we need some research—as to what we might be able to save in harvesting costs that would mean more dollars to the farmer and less expenditure of energy in getting our future energy source to the market?

Mr. TSAO. Yes. The cost of $30 per ton is, of course, based upon rather crude estimate. Out of the $30, $5 is to replace the loss of a fertilizer value when we take a ton of dry cornstalk off the land.

The land loses that much fertilizer. But the cornstalk does contain nitrogen, potash, and phosphate. Removal of a ton of that matter, we lose about $5 of fertilizer. So that has to be replaced.

Then we estimate in the area of Indiana or Illinois, in those areas which have corn and soybeans, if we draw a circle of about 30 miles radius, then in that 50-mile radius an area there would be about 16 million tons of cornstalk or dry crop residues available every year, and if we figure we can remove perhaps one-third of it, about 5 million tons can be made available for processing into alcohol.

So this means we need a delivery of up to about 50 miles distance. So with that in mind, we come up with additional $8 for trucking, just to go from the corn to the essential processing plant, which is located, say, every 100-mile distance.

Then we also include in our estimate about $5 to $7 for the additional investment in the machinery in order to collect the materials off the land.

In toto, we figure about $20 of actual costs, and so the reason we come up with $30 is we have to have additional time the farmer has to spend on collecting such materials and some additional cost. So that is how we come out with the $30 per ton figure.

Senator Bayh. So the $30 is net?

Mr. TSAO. $30 is the total payment, including that $20 for cost. So the net profit is $10 per ton.

Senator BAYH. Do you have anymore thoughts on that?

Mr. BURROWS. Yes. I would just like to add that I would agree with him completely in discussions with several farmers in our area of Illinois just asking them what would you have to have for a ton of cornstalks if you were to go to the extra trouble of gathering them and transporting them to someplace much as you do your grain and transporting it to the country elevator. They are talking in terms of about $30 a ton is what they would like to have. This can vary considerably over the country, however.

Another interesting thing that might have bearing here is that the transport costs of this kind of material is much the same in the country over. I am not sure why this should be, but hay costs to transport in the Midwest about, say, 9.8 cents per ton mile. Wood chips can be transported in the Pacific Northwest for 10 cents a ton mile; just about equivalent number or so.

I think we can get some fairly good estimates on the cost of transporting.

And you didn't let me answer your other question. I think it is certainly possible to make a machine—in fact, a machine has been made that I am aware of at least that harvests the grain and puts the other material in another bucket, so to speak. It does one job of harvesting.

Senator BAYH. I am familiar with that machine. The new component would be not only to put it in another hopper but to put it in a form that would be as compact and heavy and efficient to handle as possible. I don't think existing machinery does that.

Mr. BURROWS. That is right. Another thing that Purdue University in the agriculture engineering department has proposed is the collection of the cobs along with the grain and then using the energy that is in the cobs to both—well, to dry the grain and to transport the material within the materials handling process.

Per-Gallon Cost Of Alcohol Fuel

Senator BAYH. Dr. Tsao, let me in your price of 74 cents a gallon or 40 cents a gallon where you could use the stalks, as I recall, where it would be——

Mr. TSAO. If the raw material is of no cost, that is where I get 74 cents a gallon.

Senator BAYH. 74 cents a gallon, is that figure arrived at by using the price of corn at $1.50 cents a bushel?

Mr. TSAO. No. This is at the price of $30 per ton of the dry corn stalk. This is the basis of that, 74 cents per gallon.

Senator BAYH. I would hate for us to get a renewable waste program started in this country that would force the farmers to have to rely on $1.50 a bushel. We have real problems with that. But your figure of 74 cents a gallon is based on cellulose stalks.

Mr. TSAO. That is right. This is the waste material now, just lying on the land, not producing any dollars for the farmers. It may provide some fertilizer, more or less. But we are talking about making use of that, get some additional use.

Refining Of Process

Senator BAYH. Do you feel from where you are now that it is possible to further refine this process, both from the standpoint of dollar and cents efficiency as well as from Btu consumption of energy?

Mr. TSAO. Yes. Not so much on the dollar/cents, because the figure is probably as low as it can be, but I think we can save on the energy, as I pointed out earlier. When we stir the alcohol to distill the alcohol to make the 195-proof alcohol in order to put in automobiles, we just use the most, the very old technology of today.

We heat it up, distill and vaporize the alcohol and condense it, and when the alcohol vapor condenses it gives out heat. As of now, that heat is totally wasted. Nobody is using it. Based upon that kind of condition, that is why we start to hear talk about we are putting more in getting energy than getting out of it.

But people are studying the desalinization of seawater, they have done a tremendous amount of research on how to make full use of that energy from condensing vapor into liquid and reutilizing this energy over and over again. That is where we can do a lot more work to get much better energy efficiency.

Initial Cost Of Alcohol Research Facility

Senator BAYH. You mentioned the costs involved, or the next step or steps of research that you thought would be helpful.

What would it cost to construct a plant?

Mr. TSAO. Yes. What we need now, we have, I would say, sufficient financial support for 2 more years of laboratory research which perhaps will allow us to check out the various kinds of solvents.

As you know, I didn't specify which solvent. There is more than one solvent. We can check them out. But there is research which simply cannot be meaningfully done on a laboratory scale; for instance, how we best dissolve the cellulose with the solvent.

In a laboratory, we can do it with a glass. That doesn't tell us anything of the larger scale, the sufficient energy at least cost. So we need a larger scale to do that kind of research.

So this is the reason I put down as a need of ours $1½ million for 2 more year's work at a pilot plant scale. After that we should have all the necessary data to design the production factory.

If the past experience means anything, usually it takes about 2 to 3 years to build a production factory. That means if we have the funds available, if we have the people willing to invest, in 5 years' time we can have alcohol from the cellulose on the production scale in this country.

Federal Role In Alcohol Fuel Research

Senator BAYH. You gentlemen have all been very kind. I would like to ask, if I might, all of you, particularly Mr. Burrows, what we can do. Just think about it and let us have your thoughts after you have had a chance to think about it. All of you. What we can do in the way of Federal enticements and persuasion to get the private sector involved?

Dr. Burrows, you pointed out some of the real problems that an industry is going to have, that a company is going to have, before it is going to make the changes that would be necessary to get alcohol fuels into our energy mix. What environment do we have to create, what incentives do we have to offer, to increase the willingness, the voluntary participation necessary to get business and industry involved with the Government to get this process going? You don't need to answer that now.

Coordinating Research Council

Mr. BURROWS. I have one thing that I would like to add. There is a group called the Coordinating Research Council in New York. I can supply your staff with the name and phone number of the director.

This group is sponsored by the Society of Automotive Engineers, Motor Vehicle Manufacturers Association, to a certain degree by the American Petroleum Institute.

This is a group that has done contract work for the Government before. It is a group that has been in partnership on some projects, notably with EPA.

It is the group that does the general work on determining octane requirements for the various engines for the automotive industry.

It is a group that could very well be a vehicle for the cooperation of many, many members of the industry, both fuel suppliers and vehicle manufacturers.

And working with the Government, there wouldn't be any question of proprietary problems in the States. This is one of the problems that I see very often, and I am saying this as a scientist, not as a member of the corporation, but I find many times in the scientific work it is a little frustrating when you have worked on a project and it is very proprietary to the company that you are working for and the closer it gets to the market place, the more proprietary it seems to become.

You may have worked on this project in its beginning and yet at the end you are not supposed to know anything about it. This is one thing that the companies are very protective of and I can see the reasons for it. It directly strikes at the economic base that they have. So that has to be taken into consideration.

On the other hand, I have seen a great many things happen in agriculture, a great many advances in farm machinery that have taken place very quickly because of a clear economic advantage or economic return.

I will put it that way, rather than advantage, a clear economic return is available. I have also seen some machinery that we have developed in our research, others may have developed also in their research and brought to us, that has not gone at all because there was no market for it perhaps, or the market was very, very small, that sort of thing.

So I think if there is a clear economic return to producing an engine or a tractor or I am sure an automobile, whatever it might be in this area, to take the fuel, that that area there is no question. It will be developed.

Senator BAYH. Thank you. Do you have further thoughts on any of these questions?

Gentlemen, I hope you will let us have them so we can include them in our record. As busy as you are, I appreciate very much your taking the time to make a significant contribution to our record here this afternoon. Thank you very much.

U.S. Industry Public Witness Panel

STATEMENT OF SERGE GRATCH, DIRECTOR CHEMICAL SCIENCES LABORATORY, FORD MOTOR CO., DEARBORN, MICH.

ACCOMPANIED BY:

JOSEPH M. COLUCCI, DIRECTOR OF FUELS AND LUBRICANTS, GENERAL MOTORS RESEARCH LABORATORIES, DETROIT, MICH.

JOHN J. WISE, VICE PRESIDENT, MOBIL RESEARCH AND DEVELOPMENT CORP.

Senator BAYH. Our next panel of witnesses involve those who wear different hats in the industrial segment of our nation's economy. Dr. Gratch, the director of chemical sciences laboratory at Ford Motor; Joseph Colucci, director of fuels and lubricants of General Motors research laboratories; and John Wise, vice president of Mobile research.

Gentlemen, we appreciate your being with us and I apologize for the lateness of the hour. It is just that we have all been too exuberant in approaching this problem.

We made an error in trying to get too much done in 1 day. I appreciate your taking the time to bear with us. Why don't you proceed as you see fit?

Ford Motor Co.

Statement Of Serge Gratch

Mr. GRATCH. Senator Bayh, my name is Gratch, director of chemical sciences laboratory, Ford Motor Co.

I welcome the opportunity to testify on the important issue of the use of alcohol as a transportation fuel. In view of the hour, I will abbreviate my remarks, but I would like to make my full statement available for the record.

Senator Bayh. We will certainly do that.

[The statement follows:]

Prepared Statement Of Dr. Serge Gratch
Director, Chemical Sciences Laboratory Ford Motor Company

Senator Bayh, Committe members and guests. My name is Serge Gratch. I am Director of the Chemical Sciences Laboratory, Research, Ford Motor Company. I welcome the opportunity to testify on the important issue of the use of alcohol as a transportation fuel. Ford Motor Company supports any constructive governmental action aimed at expanding the supply of fuels, particularly those which can be obtained from abundant resources, and especially those from renewable resources.

With this objective of expanding the supply of fuels in mind, we have been investigating the use of alcohol in passenger car fuels for some time. Some of this work was done in cooperation with petroleum companies and other automotive companies through the Inter-Industry Emission Control Program. Our studies have emphasized methanol (methyl alcohol), which, in the United States—according to ERDA studies—appears to be economically more attractive when produced from coal than ethanol (ethyl alcohol) produced from grain or other agricultural products. However, we have also studied ethanol, because of its use in Brazil. Some of our work is illustrated by the attached paper which we presented at the recent International Symposium on Alcohol Fuel Technology at Wolfsburg, Federal Republic of Germany.

I will not discuss the economics associated with the use of alcohol. Rather, I will limit my remarks to the technical aspects of utilizing alcohol as a motor vehicle fuel, with particular emphasis on the results of our own experimental investigations.

From these studies, we have identified a number of advantages and disadvantages inherent in the use of alcohol and alcohol-gasoline blends. While none of the disadvantages appears insurmountable—at least from the point of view of vehicle requirements and driver satisfaction—they are real and must be considered seriously before reaching any decision to implement widespread use of alcohol as an automotive fuel. I shall briefly review our studies.

Ford Programs

Our methanol studies have included basic combustion studies and investigations of driveability, fuel economy, and emissions—with and without vehicle modifications. Extensive investigations have been made of material compatibility and phase separation. These studies have been carried out with blends and with neat (100%) methanol and have included both conventional engines and Ford's programmed combustion engine (PROCO). We have also experimented with a dual fuel system that uses both alcohol and gasoline, storing them in separate tanks and blending them in the carburetor as needed.

The studies with ethanol have been limited so far to an investigation of emissions and fuel economy with a Corcel, one of our Brazilian products, using 20% and 80% ethanol blends as well as neat ethanol. Carburetor modifications were made as needed. These studies are continuing.

A brief summary of the results of these studies follows.

Ethanol-Gasoline Blends

With additions of up to 10% ethanol, the problems appear minimal. Interpolation of the data which we have obtained on a Corcel indicates minor emission, fuel economy, and performance effects, although we must still confirm this finding for a range of engine-vehicle combinations. Also, it will be necessary to check the lists of materials used in the fuel systems of all vehicles to make sure that none of them are attacked by ethanol.

With higher ethanol concentrations, of the order of 20%, it may be necessary to make carburetor adjustments to avoid an excessively lean operation. Such an adjustment probably would not be necessary, at least up to 20% ethanol, with older cars which had a rich calibration, or with future cars which will utilize closed loop fuel air ratio control. However, carburetor adjustments probably will be necessary with most recent vehicles, in order to avoid excessive exhaust emissions. For instance, when the previously mentioned Corcel was tested with a 20% ethanol-80% gasoline blend, the emissions of oxides of nitrogen—measured on the Federal Test Procedure (the familiar CVS test)—increased to 2.39 gm/mi, 42% more than the average value of 1.68 gm/mi obtained on the same test when using gasoline alone as the fuel.

If the concentration of ethanol is increased significantly above 20%, the problems become more serious. At these higher concentrations, carburetor adjustment will be required in most cases. In addition, some changes will be required to achieve satisfactory cold starts. This point will be discussed specifically for the case of neat ethanol. Similarly, at these higher concentrations our data indicate a significant increase in the emission of aldehydes. This is highly undesirable since aldehydes are irritants, have an unpleasant odor, and are very reactive in photochemical smog formation. If it is found that such an increase in aldehyde emissions is unacceptable on the basis of air quality considerations, it will be necessary to provide additional controls. Oxidation catalysts will control these pollutants, but only after they are hot; thus aldehyde emissions during cold start represents an unsolved problem.

There also will be a need to keep gasoline distribution systems drier. The presence of alcohol increases the tendency for water pickup by the gasoline. This in turn aggravates corrosion. Moreover, if the water pickup is excessive, separation of the alcohol from the gasoline will occur at low temperatures. There have been efforts to prevent water pickup by gasoline in the past—for instance, to prevent carburetor icing—but these efforts have been relatively unsuccessful. The task will be much more difficult with alcohol present in the gasoline.

It will be necessary also to check the materials used in the fuel systems in the cars as well as in the gasoline distribution system to ensure that all the materials are compatible with ethanol. Experience in Brazil to date suggests that probably the material problems would not be major.

Methanol-Gasoline Blends

Most of the problems with methanol are similar to those with ethanol, except that approximately the same effects are produced by half as much methanol as ethanol.

Methanol, however, causes much more severe phase separation and driveability problems. The tendency to phase separation depends on the gasoline used, but is sufficiently severe with *all* gasolines that, except in tropical and subtropical climates, it will be necessary to employ additives to prevent it.

In the case of methanol we have found that, in addition to the cold start problems mentioned earlier for ethanol, there are also vapor lock problems. We are not sure whether or not these will be sufficiently severe to require some fuel system modifications for their prevention.

It appears that methanol has more severe material compatibility problems. For instance, we observed severe corrosion problems at the connection of a fuel level indicator with a gas tank when methanol was used. The petroleum industry may face a more serious material compatibility problem since methanol has been found to weaken significantly the fiberglass-reinforced plastic laminates used to line underground gasoline storage tanks. Moreover, these laminates contaminate methanol-gasoline blends. It is not yet clear whether or not this problem is sufficiently severe to require replacement of affected underground tanks.

Some of these problems can be eliminated by using the previously mentioned dual-fuel concept, wherein alcohol and gasoline are stored in separate tanks and are blended in the carburetor. We have operated two cars utilizing this concept for 50,000 miles without any major problems, except for some corrosion, which could have been prevented by material substitution. As indicated in the attached paper, this approach provides some fuel economy and performance advantages, but had a serious drawback: the emissions of hydrocarbons are increased significantly: 1,66 gm/mi, compared to 0.76 gm/mi emitted by a similar vehicle operated on gasoline only. We have experienced similar emission increases with conventional cars operated on methanol-gasoline blends.

As mentioned earlier, some of the problems with alcohol-gasoline blends—and this applies both to methanol and ethanol—can be alleviated, at least in part, by vehicle modifications. However, this leaves unsolved the problem of the more than one hundred million vehicles already in the field. Moreover, if the new cars were modified to operate well, without higher emissions, on alcohol-gasoline blends, it would be expected that they would be less than satisfactory when operated on gasoline only. This lack of interchangeability presents a major problem for the phase-in of alcohol-gasoline blends.

Neat Methanol and Ethanol

The main problem with neat methanol and ethanol is the inability to start conventional engines cold when using these fuels. Possible solutions include intake manifold heating, fuel injection, and addition of volatile additives (e.g. butane) to the fuels. More research is required to establish the effectiveness of these approaches and to determine their relative advantages and disadvantages. For instance, it has been reported in the literature than conventional fuel injection is not sufficient to allow satisfactory cold starting when the ambient temperature is below 40°F.

Our data indicate that the increase in aldehyde emissions is quite noticeable with neat methanol or ethanol. For instance in

CVS tests with a Corcel fueled with ethanol we have observed more than a fourfold increase in aldehyde emissions, for 1.7 to 7.6 parts per million. More work is required to resolve this issue.

Other problems with neat alcohols (or with blends containing very high fractions of alcohol) are that carburetors will definitely require recalibration and that, because of the lower heat of combustion of the alcohols, larger fuel tanks will be required to achieve the same driving range between refueling stops; the increase will be roughly a factor of two for methanol and one and a half for ethanol. This is a very unattractive feature in view of the current need to minimize vehicle size in order to attain the mandated fuel economy standards.

Conclusions and Recommendations

In my judgment none of the problems outlined above are insurmountable. In general, Ford Motor Company supports consideration of possible governmental actions which would provide incentive for the development of alternative energy sources. However, we would urge that such proposals provide sufficient flexibility to exploit whichever technology is most effective in meeting the overall energy goals of the country.

For example, we believe that equal consideration should be given to the use of *any* alcohol, be it methanol, ethanol, or other. Which alcohol should be favored is not yet known.

Similarly, if a governmental incentive is found desirable, it should apply equally to the direct use of alcohol and to the use of alcohol after conversion to gasoline. Recent reports indicate that alcohol can be converted to high quality gasoline efficiently and at low cost.

Any proposed incentives should also apply equally to the use of alcohol produced from sources such as grain, sugar cane, and beets as well as to liquid fuels produced from cellulose and other materials present in agricultural and forestry wastes, and from any other currently unexploited source.

Finally, the incentives should apply equally to fuels produced from solar energy, both by the agricultural product route and by other means such as non-biological photochemical reactions.

This flexibility is essential in view of the unsolved problems and open issues connected with the prospective use of alcohols and alcohol-gasoline blends as automotive fuels, including, for instance:

—potential adverse impact on air quality;

—potential driveability problems;

—potential corrosion and other material compatibility problems.

As I stated in my opening remarks, although probably not insurmountable, these problems are sufficiently serious to demand careful consideration.

In closing, I express my support for the examination of possible means to encourage the production of liquid fuels from sources other than petroleum, but I urge you to broaden any legislative proposals so as to apply not only to gasohol, but also to all other liquid fuels which could be made from abundant domestic resources.

Utilization of Methanol as an Automotive Fuel - A Report from IIEC-2, the Inter-Industry Emission Control Program

R.E. Baker, J.A. Harrington
Ford Motor Company

J.H. Baudino, L.C. Copeland
Atlantic Richfield Company

W.J. Koehl, L.J. McCabe
Mobil Research and Development Corp.

W.T. Wotring
The Standard Oil Company (Ohio)

Introduction

Potential advantages of methanol, relative to gasoline, are most evident when it is used in the neat form. Those advantages include increased power, reduced specific energy consumption, an extended lean misfire limit, reduced NO_x emissions, and increased fuel octane quality (1). An important source of those advantages is the high latent heat of vaporization which also introduced a serious disadvantage. The cooling that results when methanol vaporizes in a conventional intake manifold limits fuel vaporization too much for engine starting below 10 to 15°C or for acceptable operation after starting unless supplemental heating is applied (2). But, as will be shown, this heating reduces the benefits attainable.

Blending methanol in gasoline minimizes the problems associated with latent heat of vaporization but also tends to minimize any advantages in power, specific energy consumption of NO_x emissions. Some of these effects will be illustrated here. Additional problems associated with blends are greatly increased vapor pressure and facile extraction of methanol by water present inadvertently in fuel handling systems (2,3).

A third approach to exploiting some of the advantages of methanol while avoiding the problems is demonstrated here in a dual-fuel system in which methanol is added to the charge at the carburetor only when required for knock suppression and power.

Common to all three approaches to using methenol are problems of compatibility with materials used in fabricating fuel handling systems. This concern will be illustrated with data for laminates used in gasoline storage tanks.

All of these advantages and disadvantages must be weighed in selecting the best approach for using methanol in any given situation.

Heat Methanol

1. Heat of Vaporization Effects

Intake Temperature Depression. The high latent heat of vaporization of methanol causes significant cooling of intake manifold and/or in-cylinder gases. As shown in Table B-6 substitution of methanol for gasoline in a carbureted single cylinder engine (a Waukesha variable compression ratio engine (1), depresses intake mixture temperatures by about 30°C. Air entering the carburetor at 38°C is cooled to 28°C at the intake port with gasoline and to –2°C with methanol. The temperature decrease with methanol translates to vaporization of about 20% of the fuel while the much smaller temperature decrease with gasoline reflects roughly 40% vaporization. With methanol the 80% of the fuel left as a liquid acts as an in-cylinder heat sink during compression. The combined effects of intake and in-cylinder cooling contribute to an increase in peak power and to a reduction in knocking tendency and NO_x emissions. The cooling of in-cylinder gases can also have a negative effect by slowing down the combustion process and promoting quench layer growth and increased unburned fuel emissions. With carbureted multi-cylinder engines, manifold temperature reductions and fuel vaporization levels are about equal to those for single cylinder operation, thus mixture distribution, mixture quality and the related driveability characteristics are adversely affected.

Prevaporized Methanol. Utilization of methanol as a neat fuel will require measures to improve mixture distribution. Measures usually considered include increasing manifold heat and use fuel injection. When manifold heat is increased, some of the benefits of methanol use will be compromised. Studies with the same carburetted single single cylinder engine showed that the NO_x emission advantage over gasoline is greatly reduced when

Table B-6. Intake Mixture Temperatures, °C, of a Carbureted Single Cylinder Engine With Inlet Air Temp = 38°C.*

Fuel/Air Equiv. Ratio	Methanol	Gasoline
1.2	−2	26
1.0	−2	28
0.8	−2	32

*1500 RPM, 700 mm Hg manifold pressure

methanol is introduced as a vapor. As shown in Fig. 1, NO_x emission with methanol vapor is much higher than with the liquid and in fact is almost equal to that with gasoline. The HC emission with methanol vapor is reduced relative to that with the liquid at fuel rich mixture ratios, Fig. B-14. However, the resultant HC concentrations with methanol vapor are still equal to or higher than with gasoline at lean ratios.

Other effects found for vapor versus liquid methanol operation are a slight extension of lean limit and an expected increase in burn rate; CO emission is unaffected, however engine power at a constant value of air flow, fuel flow and spark timing increases slightly and this improves fuel consumption.

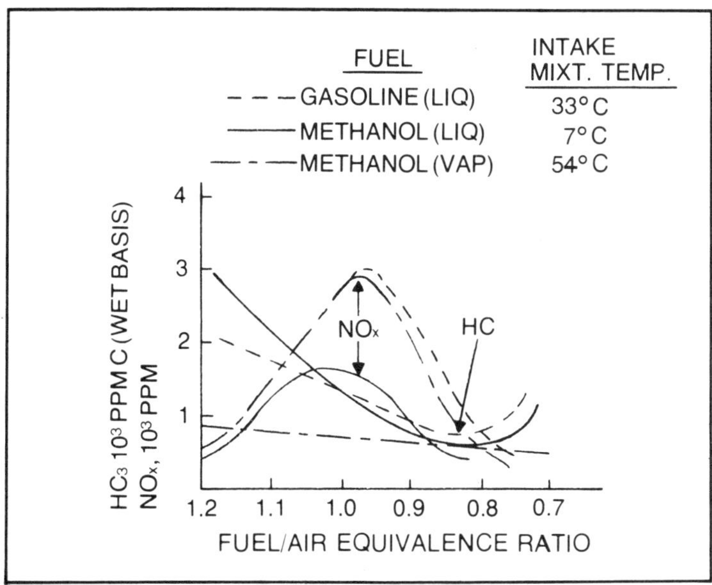

Fig. B-14. Effect of Fuel on HC and NO_x emmissions from a single cylinder engine (1000 rpm, 15° BTDC, 700mm Hg manifold pressure).

The single cylinder engine results should illustrate the effects that are likely in multi-cylinder engine operation. In the limiting case of fully vaporized methanol, a uniform mixture distribution should be obtained and this should be reflected in driveability and lean limit operation. The directional effect on emissions would be about as indicated with the single cylinder engine except where catalytic exhaust treatment is required. Exhaust gas temperatures with methanol are generally lower than with gasoline and if exhaust heat is to be utilized for fuel vaporization, maintenance of temperatures sufficient to assure catalyst function could be a problem. The actual level of fuel vaporization required for satisfactory driveability in vehicles is expected to be less than 100 percent and therefore the effects would be smaller than indicated above. The added manifold heat would improve overall driveability at the expense of some power and a possible NO_x increase, if the extended lean limit is not exploited to reduce NO_x. Also, cold start behavior could continue to be a problem with carburetted engines.

2. Exhaust Emission Species

Analytical Procedures. Analytical procedures presently used for emissions certification were developed with the presumption that the fuel would be gasoline. These procedures do not account adequately for all important exhaust species with methanol fuels, since the exhaust contains unburned alcohol and aldehydes, to which FID instruments to measure HC emissions do not respond fully.

For this study a computer controlled gas chroatograph/mass spectrometer procedure was used to measure methanol and various hydrocarbons. Aldehydes were not eluted through the GC column, so a colorimetric technique for total aldehydes was developed using the MBTH reagent.

Engine Tests. Engine dynamometer tests were conducted on a 1975 6.6L V-8 engine using unleaded gasoline (Indolene clear), a blend of 20% methanol in Indolene and neat methanol. Methanol and aldehyde emissions measured at 1000 and 3000 rpm and 25% of full torque are shown in Table 2. The fuel-air equivalence ratio was maintained at 0.95. Spark timing was held at MBT, and EGR was not used. Similar trends were found at WOT.

Both methanol and aldehydes are present in small amounts when using gasoline and increase non-linearly at each speed with the methanol content of the fuel.

Table B-7 also shows the trends of methanol and aldehydes with increasing engine speed. The magnitude of the reduction between the two speeds is a measure of the reactivity of the

Table B-7. Sensitivity of Unburned Fuel Emissions to Engine Speed.

Methanol In Fuel, %	Engine Speed, rpm	Exhaust	
		Methanol, ppm	Aldehyde, ppm
0	1000	Trace	60
	3000	Trace	25
20	1000	50	80
	3000	20	50
100	1000	900	300
	3000	30	30

species since exhaust temperature increases with speed. Total HC emissions with gasoline alone at 1000 rpm were about equal to the methanol emissions of methanol alone. At 3000 rpm the HC emissions were reduced only about 70% while the methanol emissions were reduced 95%. Similarly the high aldehyde concentration with neat methanol is reduced about 90% at the higher speed. Hence both methanol and aldehyde are more easily oxidized than HC.

Aldehyde emissions are known to increase in vehicle tests as well, with both neat methanol and with blends (2); however, exhaust oxidation catalysts have been observed to reduce them to very low levels.

Methanol Blends

1. 80,000 km. Road Test

Much has been written about the performance of methanol-gasoline blends in vehicles as measured by driveability, emissions, and fuel economy; but little has been reported on the long-term durability of vehicles using these fuels. This study was initiated to obtain such data in road driving under conditions of rapid mileage accumulation. As a further departure from most published work in which blends were made from relatively pure, chemical grade methanol, a mixture of alcohols simulating the expected product of a plant making fuel grade methanol (4), was used.

Test Program. The performance of a gasoline blend containing 20% of mixed alcohols ("M-fuel") was compared with that of a commercial unleaded gasoline in an 80,000 km. road test. The composition of the alcohol mixture was 65 vol. % methanol, 7% ethanol, 21% 2-methyl propanol, and 7% n-hexanol.

Table B-8 summarizes the properties of the fuels. The M-fuel was blended from a gasoline of reduced Reid vapor pressure so that the expected increase on adding the alcohol mixture would cause its RVP to approach that of the reference fuel. Both M-fuel and the reference fuel contained an intake system cleanliness additive package consisting of a polybutene amine dispersant and a carrier oil.

Test cars were full-sized 1975 models with 6.6L V-8 engines. Air fuel ratio for the cars run on M-fuel was adjusted to approximately the same equivalence ratio as in the cars run on reference fuel by increasing the size of the main carburetor jets. This was done to avoid the leaning-out effect encountered when fuels containing alcohols are run in vehicles with standard carburetion. All cars had production exhaust emission control systems, including an oxidation catalyst on one bank of cylinders. Two cars were run on each test fuel.

Mileage was accumulated by operating the cars on interstate highways as part of a fleet to assist stranded motorists. Cars were run 16 hours per day with an average accumulation of 800 km. Maximum speed was 88 kph with considerable idle and shut down periods. Motor oil was a 5W-40 grade of SE quality with oil drains at 13,000 km. intervals.

Exhaust Emissions. During the 80,000 km. test, tailpipe and catalyst feedgas emissions were monitored at 13,000 and 26,000 km. intervals respectively using the U.S. Federal Test Procedure. Examining overall trends revealed no effect of the mixed alcohol fuel on CO emissions because of the adjustment in air-fuel ratio of the cars using the fuel. Likewise, NO_x emissions were found to be slightly lower with M-fuel, as could be expected from the lowered peak combustion temperature of alcohols; but in

Table B-8. Test Fuel Inspections.

	"M-Fuel"	Reference Fuel
RVP, bar	0.65	0.72
Distillation, °C		
10% Evap.	48	44
30% Evap.	64	77
50% Evap.	85	106
90% Evap.	162	176
EP	209	220
Research Octane	99.2	92.0
Motor Octane	86.4	84.0
% Aromatics	24.5	26.0
% Olefins	4.5	3.3
% Saturates	51.0	70.7

Table B-9. Fuel Economy, L/1000 km.

M-Fuel		Reference	
Car 1	19.18	Car 3	17.12
Car 2	18.92	Car 4	17.42
Avg.	19.05	Avg.	17.27

view of test repeatibility the difference may not be significant. Also overall hydrocarbon emissions measured by an FID instrument appeared slightly higher with M-fuel both at the tailpipe and in the catalyst feedgas. Again test variability obscures the significance of this trend, but because it also appeared in a similar road test conducted a year earlier (3), and because the FID method underestimates the difference (by underestimating unburned methanol) further study may be warranted.

Fuel Economy—there was little variation in fuel economy calculated from driver records of gasoline usage over the 80,000 km. test period. The average values for each car, shown in Table B-9 illustrate the difference between the two test fuels. Fuel consumption with M-fuel increased 1.78 L/100 km, or 10.3% over that with the reference fuel. This difference approximates that expected from the reduction in heating value caused by the mixed alcohols. This reduction in fuel economy was similar to that observed in the 1974 road test (3).

Motor Oil Economy and Used Oil Analysis. Motor oil economy was higher in the M-fuel cars than in the reference cars, as shown in Table B-10. This result was similar to that obtained in the 1974 program where oil economy was higher in cars run on fuel containing 9% methanol and 1% 2-propanol than in cars run on straight gasoline (6500 vs. 4600 km./L, respectively) (3).

Oil economy was relatively constant over the 80,000 km. test except in one M-fuel car in which it decreased from about 8500 km./L initially to about 3400 km./L at 64,000 km.

Table B-10. Motor Oil Economy, km./L.

1975 Program			
M-Fuel		Reference	
Car 1	5685	Car 3	3696
Car 2	9773	Car 4	5996
	7729		4846

Table B-11. Average Used Oil Inspections.

	M-Fuel		Reference	
	Car 1	Car 2	Car 3	Car 4
Carbon Residue, %	1.524	1.455	1.943	1.835
Pentane Insolubles, %	0.038	0.032	0.047	0.064
Viscosities*				
cs @ 99°C(210°F)	18.05	16.38	15.74	16.10
cs @ 38°C(100°F)	126.08	116.43	104.74	108.47
poise @ – 18°C(0°F)	39.5	34.6	26.7	25.7

*Fresh oil viscosities were 1.51 cs @ 99°C, 76.1 @ 38°C, and 11.8 @ – 18°C.

Used oils were tested for acid number, base number, carbon residue, pentane insolubles, and viscosities at each drain period. Acid and base numbers showed no significant differences. Pentane insolubles and carbon residues were lower for M-fuel cars, as shown in Table B-11. Oils from those cars also had significantly higher viscosities than the oils from the reference cars, despite the fact that carbon residue and pentane insolubles showed less contamination.

The difference in oil viscosities between the base gasoline and alcohol blend was not as great in the 1974 program. There was no significant effect at 99°C, but –18°C viscosities were 42.6 poise for base fuel and 45.6 poise for the alcohol blend. Lower alcohol concentrations and the use of only methanol and isopropanol in 1974 may account for the smaller effect on oil viscosity. However, increased oil viscosity and reduced consumption with fuels containing 9% methanol and 1% 2-propanol have been found in other makes of cars in similar programs with a variety of motor oils. Table B-12 shows these results.

Table B-12. Additional Examples of Effects of Alcohols on Motor Oil Viscosity and Consumption.

	Pairs Of Cars	Visc., cs @ 99°C		Visc., Poise @ – 18°C		Oil Economy, km./L	
		Base Fuel	Alcohol Blend	Base Fuel	Alcohol Blend	Base Fuel	Alcohol Blend
Make A	2	22.29	22.68	21.2	22.0	4494	7193
Make B	2	26.14	29.92	24.3	32.8	3759	4161
Make C	1	16.98	17.45	16.5	21.6	6632	7051
	Avg.	21.80	23.35	20.7	25.5	4692	6135

Certainly additional data are required to evaluate the effects of alcohol on oil consumption and oil thickening; and, these potential detrimental effects should receive careful attention in any consideration of the commercial use of alcohols as fuel components.

Engine Deposits and Wear - Overall engine cleanliness was very good for all four cars, but there were several points of difference between the two sets of cars. The M-fuel cars formed slightly less varnish on all rating surfaces except the rocker arm covers and considerably less deposit on intake valves than the reference cars. Carburetors on all four cars were very clean, but there was slightly more deposit on the throttle bodies of the M-fuel cars. This may indicate that the carburetor detergent used was less effective in the fuel containing alcohols.

No evidence of unusual wear was found on camshaft lobes, tappet crowns, valve tips, valve guide bores, piston pins or top compression ring gaps, or in the iron and copper contents of the used oils.

2. Driveability Tests

Driveability has been reported to be impaired somewhat in cars fueled with methanol blends (2,5). In the 80,000 km. test described above, no driveability differences were noted, presumably because the carburetors of the M-fuel cars had been adjusted for the mixture leaning effect of alcohols. A brief driveability study was conducted with 10 and 20% "alcohol blends" in an unadjusted car. A second variable in this test was base gasoline volatility to distinguish between mixture leaning and blend volatility effects on driveability.

Test Program - The test vehicle was a full-sized 1974 U.S. model equipped with an automatic transmission and a 6.6L 2-V, V-8 engine calibrated to the 1974 49-state emission standards. Chassis dynamometer inertia loading totaled 2130 kg. for this vehicle.

The test fuels consisted of three additive-free, unleaded gasolines, representing low, intermediate, and high volatility (Fuels 11, 12 and 13) as defined by their 50% and 90% evaporation points; and blends of these fuels with 10 and 20 vol. % of alcohol—a 4:1 volume mixture of methanol and t-butanol. This mixture was chosen to improve the low temperature phase stability and water tolerance of the alcohol-gasoline blends (3). Distillation characteristics of the base fuels and 20% vol. blends are summarized in Table B-13.

Table B-13. Distillation Characteristics of the Test Fuels.

Alcohol*, Vol. %	11		12		13	
	0	20	0	20	0	20
Distillation, °C (ASTM D-86)						
10% Evap.	51	46	50	44	46	45
50% Evap.	113	80	94	60	74	57
90% Evap.	175	173	158	153	105	105

*Methanol/t-butanol in 4:1 volume ratio.

Driveability testing was conducted on a chassis dynamometer at an ambient temperature of 21 ± 2°C. The procedure consisted of the CRC Intermediate Temperature Driveability Schedule (6) for cold start and cold driveaway performance plus a series of constant speed driving modes for warm-up followed by a series of maneuvers to determine hot driveability (fully warmed-up) characteristics.

Consumer driveability testing was accomplished by rotating the vehicle through six drivers for use in their normal "home-to-work" driving, while rotating the fuels in random order. The drivers noted the frequency and severity of malfunctions related to driveability.

Driveability Results. Driveability performance measured on the chassis dynamometer is summarized in Fig. B-15 as a function of base fuel volatility and alcohol content. Driveability is expressed as total demerits assigned and weighted for various malfunction modes according to the CRC procedure. These data show that driveability at 21°C depreciates with decreasing base fuel volatility or increasing alcohol content. Driveability of Fuel 11 blended with 20 vol. % alcohol is considered unsatisfactory. These results are consistent with data reported by Crowley et. al. (3).

Table B-14 summarizes results of the consumer driveability test. These data are average results from six drivers. The correlation between demerits in the consumer and chassis dynamometer tests was satisfactory and had a correlation coefficient which implied that chassis rating is a valid method of measuring driveability performance. It is also apparent from this study that the consumer would detect increased incidence and severity of malfunctions in vehicles which were fueled with alcohol blends without being modified for operation on these fuels.

Driveability results on the chassis dynamometer were analyzed for demerit distribution according to malfunction. De-

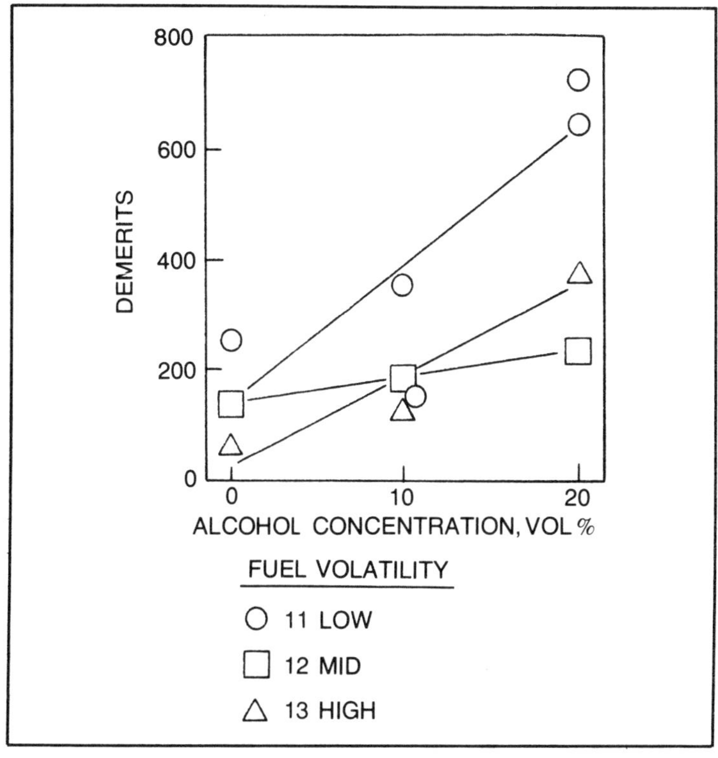

Fig. B-15. Effects of alcohol (4:1 methanol + t-butanol) and base fuel volatility on vehicle driveability at 21°C.

merits for hesitation, stumble, surge, and stretchiness generally accounted for 70% or more of the total. Stalls were a significant contributor only with the low volatility fuel. Idle roughness and backfire demerits were only minor contributors.

Further evidence that the driveability degradation seen with increasing methanol concentration was caused by mixture enleanment rather than fuel volatility changes was found in emissions data. Hydrocarbon emissions increased slightly while CO and NO_x

Table B-14. Consumer Driveability Evaluation of a 1974 Automobile.

Alcohol Vol. %	Average Demerits	
	Fuel 12	Fuel 13
0	27	7
10	85	-
20	108	-

emissions decreased markedly as the alcohol concentration was increased. These results are as expected with mixtures leaner than the stoichiometric A/F.

Dual-Fuel Concept

Besides driveability concerns discussed above, two other potentially more serious problems of methanol blends are vapor locking, because of the large vapor pressure increase found on adding methanol to gasoline, and phase separation, because of the ready solubility of methanol in inadvertent water in fuel handling systems. All of these problems can be avoided while still realizing the octane and power benefits of methanol, if the methanol is supplied to the engine through a separate methanol fuel system. The higher octane value of the blend formed in the intake system can be used to improve fuel economy by increasing compression ratio. This provides higher engine power which can be used to improve fuel economy further by reducing engine size to obtain equal vehicle performance.

Test Program. Two 1975 vehicles with 5.8L, 9.5 CR engines were used to evaluate the dual-fuel concept. The cars accumulated 80,000 km. on interstate highways along with two 6.6L, 8.0 CR vehicles which served as the baseline for both the dual-fuel cars and for the M-fuel cars reported above. The production 5.8L and 6.6L engines were identical except for stroke, so that the comparison of emissions or fuel consumption would not be compromised by design differences, such as cylinder head or block. The pistons in the 5.8L engines were changed to obtain 9.5 CR from production 8.0 CR engines. The vehicles were equipped with exhaust oxidation catalysts.

The methanol supply system is shown in Fig. B-16 This was completely separate from the gasoline system. A 30L tank was located in the trunk, and an electric pump, filter and pressure regulator were used to supply methanol to a small reservoir attached to the carburetor fuel bowl. An overflow return line permitted continuous circulation of methanol, which was found necessary to prevent vapor lock. Carbon canisters were used for collection of methanol vapor emissions.

The carburetor was modified to supply neat methanol through the power enrichment system. The power valve restrictors were enlarged to provide the same equivalence ratio as with gasoline enrichment. This provided a blend of 25% methanol in the intake manifold whenever power enrichment was required, below 125 mm Hg. manifold vacuum.

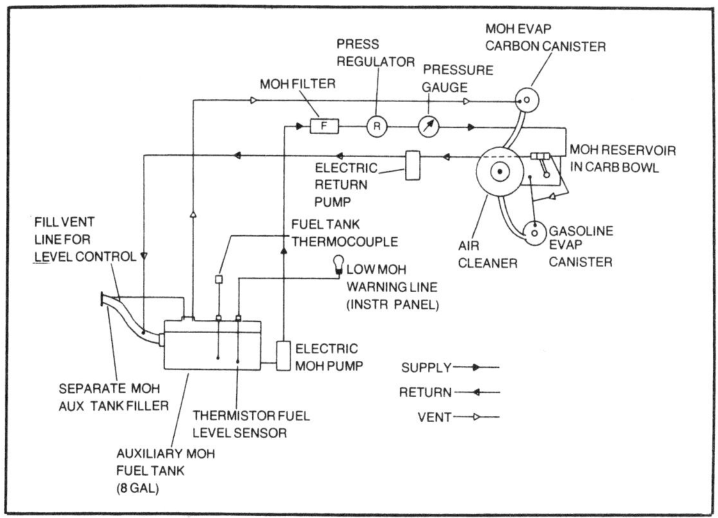

Fig. B-16. Dual-fuel vehicle: methanol fuel system schematic.

Emissions, fuel economy and octane requirements were measured at the beginning of the test program and periodically throughout the mileage accumulation. WOT performance was also determined. The same engine calibrations, namely those for 1975 49-state production 6.6L engines, were used on all vehicles to facilitate a direct comparison of engine size and compression ratio.

Results - The results summarized in Table 10 are averages for each pair of cars. The emissions are the final values at 80,000 km., while the fuel economy is the accumulated average for the highway driving. The only significant change in emissions is the large HC increase due to higher compression ratio. It is apparent that the dual-fuel concept is not appropriate if the compression ratio of the engine is limited by HC rather than octane. The emission characteristics of the methanol-gasoline blend are unimportant since very little power enrichment is required on the emissions driving cycle. Methanol consumption was also small during the mileage accumulation, which consisted primarily of cruising at 88 kph and idling.

The volume of methanol used was 2.8% of the gasoline used. This fraction would be higher for urban and suburban driving where moderate and heavy accelerations are more frequent.

The fuel economy of the dual-fuel vehicles was significantly improved, both by the higher compression ratio and the smaller engine size. The fuel consumption values in Table B-15 are in

Table B-15. Dual-Fuel Car Performance.

	Baseline Engine 6.6L, 8 CR	Dual-Fuel Engine 5.8L, 9.5 CR
1975 FTP Emissions		
Feedgas at 80,000 km.		
HC, g/km.	0.47	1.03
CO, g/km.	9.1	11.3
NO_x, g/km.	1.7	1.7
Highway Fuel Consumption		
Over 80,000 km.		
L/100 km. as Gasoline	17.3	16.2
WOT Performance	13.3	12.0
0 to 96 kph, sec.		

terms of gasoline, the methanol being converted to a volume of gasoline containing equivalent energy. In addition to the economy improvement, the dual-fuel vehicles also provided better performance as measured by acceleration times. If the engine size had been reduced further to equal performance, the economy gain would have been larger.

The stabilized octane requirements of the test vehicles, as measured by the CRC E-15 procedure, were 87 to 89 (FBRU RON)—i.e. below the 91 RON. See Table B-16.

The data indicate that fuels containing methanol may require a higher concentration of oxidation inhibitor, and that blends in the field would have to be monitored closely until more experience has been accumulated.

Gasoline From Methanol

Several engineering problems have been identified in this paper that will require solutions before methanol can be used reliabily as an automotive fuel. Another approach, which would circumvent the engineering problems, is conversion of methanol to

Table B-16. Fuel Gums From Fiberglass Laminate Compatibility Tests.

	Existent Gum, mg/100 ml		
Alcohol in Blend	None	8% Methanol + 2% t-Butanol	10% Methanol
First 3 Months			
No Laminate	3	8	12
Laminate	5	44	34
Second 3 Months			
Laminate	7	32	24

gasoline. A simple catalytic process has been demonstrated to effect this conversion (7). Using a zeolitic catalyst nearly 90% yields of product with the composition, octane number, boiling range and other specifications of high-quality gasoline can be obtained. Typical properties of the gasoline are given in Table B-17.

Conclusions

Findings reported here that are of practical concern in considering approaches to using methanol as a fuel are the following.

(1) Vaporization of neat methanol to improve startability and driveability reduces the peak power and NO_x emissions benefits associated with the high latent heat of vaporization of methanol.

(2) Aldehyde emissions increase over gasoline with both neat methanol and methanol blends, but their concentrations are reduced with increased exhaust temperatures.

(3) In an 80,000 km. test of a 20% alcohol (65% methanol) blend no significant performance or operating problems were observed; however, motor oil viscosity increased and oil consumption was reduced relative to gasoline.

(4) In both chassis dynamometer and consumer driving test of a car without carburetor adjustments, driveability depreciated with increasing methanol concentration in gasoline.

(5) A dual-fuel system was demonstrated in an 80,000 km. road test. Power enrichment with methanol along with normal operation on gasoline gave better acceleration and fuel economy but higher HC emissions from a 9.5 CR, 5.8L engine than from an 8.0 CR, 6.6L engine using gasoline alone in an identical car.

(6) Compatibility studies with a Fiberglass laminate in a 10% methanol-gasoline blend showed marked changes in the laminate. According to the tank manufacturer, the changes probably are not beyond acceptable limits for underground tanks. The tests also showed a trend toward increased gum content in the blend.

Table B-17. Typical Properties of Gasoline From Methanol.

RVP, bar	0.62
Distillation, °C	
10% Evap.	47
30% Evap.	64
50% Evap.	94
90% Evap.	159
Research Octane Number	94
Specific Gravity	0.720

(7) Other approaches such as converting methanol to gasoline exist for using any available methanol while circumventing the engineering problems of using it directly as an automotive fuel.

Acknowledgment

We want to thank J.A. Zaloudek and L.G. Pinard of Owens/Corning Fiberglass Corporation for their help in the compatibility tests.

References

1. J.A. Harrington and R.M. Pilot, "Combustion and Emission Characteristics of Methanol", SAE Paper No. 750420, Feb. 1975.
2. Alcohols - A Technical Assessment of Their Use as Fuels, American Petroleum Institute, Pub. No. 4261, July 1976.
3. A.W. Crowley, et.al., "Methanol-Gasoline Blends—Performance in Laboratory Tests and in Vehicles", SAE Paper No. 750419, Feb. 1975.
4. F. Renige and K. Zepf, Brennstoff. Chem. 35, 167 (1954).
5. W. Lee, et.al., "Volkswagen Methanol Program and the Results of Vehicle Fleet Test", Symposium on Alcohols as Alternative Fuels, Toronto, No. 1976.
6. Driveability Evaluation in Cool Weather, Coordinating Research Council, Inc. New York, 1970.
7. S.L. Meisel, et.al., "Gasoline from Methanol in One Step", Chemtech, 6, 86 (1976).

Fuel Supply Expansion

Mr. GRATCH. The Ford Motor Co. supports any constructive governmental action aimed at expanding the supply of fuels, particularly those which can be obtained from abundant resources, and especially those from renewable resources.

With this objective of expanding the supply of fuels in mind, we have been investigating the use of alcohol in passenger car fuels for some time. Some of this work was done in cooperation with petroleum companies and other automotive companies through the inter-industry emission control program.

Our studies have emphasized methanol—methyl alcohol—which, in the United States—according to ERDA studies, appears to be economically more attractive when produced from coal than ethanol—ethyl alcohol—produced from grain or other agricultural products.

However, we have also studied ethanol, because of its use in Brazil. Some of our work is illustrated by the attached paper which we presented at the recent International Symposium on Alcohol Fuel Technology at Wolfsburg, Federal Republic of Germany.

Technical Aspects of Utilizing Alcohol

I will not discuss the economics associated with the use of alcohol. Rather, I will limit my remarks to the technical aspects of utilizing alcohol as a motor vehicle fuel, with particular emphasis on the results of our own experimental investigations.

From these studies, we have identified a number of advantages and disadvantages inherent in the use of alcohol and alcohol-gasoline blends. While none of the disadvantages appears insurmountable—at least from the point of view of vehicle requirements and driver satisfaction—they are real and must be considered seriously before reaching any decision to implement widespread use of alcohol as an automotive fuel.

I shall briefly review our study.

Review Of Study

Our methanol studies have included basic combustion studies and investigations of driveability, fuel economy, and emissions—with and without vehicle modifications. Extensive investigations have been made of material compatibility and phase separation. These studies have been carried out with blends and with neat—100 percent methanol and have included both conventional engines and Ford's programed combustion engine—PROCO. We have also experimented with a dual fuel system that uses both alcohol and gasoline, storing them in separate tanks and blending them in the carburetor as needed.

The studies with ethanol have been limited so far to an investigation of emissions and fuel economy with a Corcel, one of our Brazilian products, using 20- 80-percent ethanol blends as well as neat ethanol. Carburetor modifications were made as needed. These studies are continuing. A brief summary of the results of these studies follows.

With additions to up to 10 percent ethanol, the problems appear minimal. Interpolation of the data which we have obtained on a Corcel indicates minor emission, fuel economy, and performance effects, although we must still confirm this finding for a range of engine-vehicle combinations. Also, it will be necessary to check the lists of materials used in the fuel systems of all vehicles to make sure that none of them are attacked by ethanol.

With higher ethanol concentrations of the order of 20 percent, it may be necessary to make carburetor adjustments to avoid an excessively lean operation. Such an adjustment probably would not be necessary, at least up to 20 percent ethanol, with older cars which had a rich calibration, or with future cars which will utilize closed loop fuel air ratio control.

Senator BAYH. If you would excuse me, I will have to run over to the floor. I will be right back. I have the same time schedule. If I could ask you to proceed, I will be back by the time you are through. I am embarrassed by this. These things happen.

Carburetor Adjustments

Mr. GRATCH. However, carburetor adjustments probably will be necessary with most recent vehicles, in order to avoid excessive exhaust emissions. For instance, when the previously mentioned Corcel was tested with a 20 percent ethanol-80 percent gasoline blend, the emissions of oxides of nitrogen—measured on the Federal test procedure—the familiar CVS test—increased to 2.39 gm/mi, 42 percent more than the average value of 1.68 gm/mi obtained on the same test when using gasoline alone as the fuel.

If the concentration of ethanol is increased significantly above 20 percent, the problems become more serious. At these higher concentrations, carburetor adjustment will be required in most cases. In addition, some changes will be required to achieve satisfactory cold starts. This point will be discussed specifically for the case of neat ethanol.

Similarly, at these higher concentrations, our data indicate a significant increase in the emission of aldehydes. This is highly undesirable since aldehydes are irritants, have an unpleasant odor, and are very reactive in photochemical smog formation.

If it is found that such an increase in aldehyde emissions is unacceptable on the basis of air quality considerations, it will be necessary to provide additional controls. Oxidation catalysts will control these pollutants, but only after they are hot; thus aldehyde emissions during cold start represents an unsolved problem.

Maintenance Of Dry Gasoline Distribution Systems

There also will be a need to keep gasoline distribution systems drier. The presence of alcohol increases the tendency for water pickup by the gasoline. This in turn aggravates corrosion. Moreover, if the water pickup is excessive, separation of the alcohol from the gasoline will occur at low temperatures. There have been efforts to prevent water pickup by gasoline in the past—for instance, to prevent carburetor icing—but these efforts have been relatively unsuccessful. The task will be much more difficult with alcohol present in the gasoline.

It will be necessary also to check the materials used in the fuel systems in the cars as well as in the gasoline distribution system to ensure that all the materials are compatible with ethanol. Experience in Brazil to date suggests that probably the material problems would not be major.

Problems with Methanol

Most of the problems with methanol are similar to those with ethanol, except that approximately the same effects are produced by half as much methanol as ethanol.

Methanol, however, causes much more severe phase-separation and driveability problems. The tendency to phase-separation depends on the gasoline used, but is sufficiently severe with all gasolines that, except in tropical and subtropical climates, it will be necessary to employ additives to prevent it.

In the case of methanol, we have found that, in addition to the cold start problems mentioned earlier for ethanol, there are also vapor lock problems. We are not sure whether or not these will be sufficiently severe to require some fuel system modifications for their prevention.

It appears that methanol has more severe corrosion problems at the connection of a fuel level indicator with a gas tank when methanol was used. The petroleum industry may face a more serious material compatibility problem since methanol has been found to weaken significantly the fiberglass-reinforced plastic laminates used to line underground gasoline storage tanks. Moreover, these laminates contaminate methanol-gasoline blends.

It is not yet clear whether or not this problem is sufficiently severe to require replacement of affected underground tanks.

Dual-Fuel Concept

Some of these problems can be eliminated by using the previously mentioned dual-fuel concept, wherein alcohol and gasoline are stored in separate tanks and are blended in the carburetor. We have operated two cars utilizing this concept for 50,000 miles without any major problems, except for some corrosion, which could have been prevented by material substitution.

As indicated in the attached paper, this approach provides some fuel economy and performance advantages, but had a serious drawback: the emissions of hydrocarbons are increased significantly; 104 gm/mi, compared to 0.76 gm/mi, emitted by a similar vehicle operated on gasoline only.

We have experienced similar emission increases with conventional cars operated on methanol-gasoline blends.

As mentioned earlier, some of the problems with alcohol-gasoline blends—and this applies both to methanol and ethanol—can be alleviated, at least in part, by vehicle modifications. However, this leaves unsolved the problem of the more than 100-million vehicles already in the field.

Moreover, if the new cars were modified to operate well, without higher emissions on alcohol-gasoline blends, it would be expected that they would be less than satisfactory when operated on gasoline only. This lack of interchangeability presents a major problem for the phase-in of alcohol-gasoline blends.

Problems With Neat Alcohol

The main problem with neat methanol and ethanol is the inability to start conventional engines cold when using these fuels. Possible solutions include intake manifold heating, fuel injection, and addition to volatile additives, that is, butane, to the fuels.

More research is required to establish the effectiveness of these approaches and to determine their relative advantages and disadvantages. For instance, it has been reported in the literature that conventional fuel injection is not sufficient to allow satisfactory cold starting when the ambient temperature is below 40°Fahrenheit.

Our data indicate that the increase in aldehyde emissions is quite noticeable with neat methanol or ethanol. For instance, in CVS tests with a Corcel fueled with ethanol, we have observed more than a fourfold increase in aldehyde emissions, from 1.7 to 7.6 p/m. More work is required to resolve this.

Other problems with neat alcohols—or with blends containing very high fractions of alcohol—are that carburetors will definitely require recalibration and that, because of the lower heat of combustion of the alcohols, larger fuel tanks will be required to achieve the same driving range between refueling stops; the increase will be roughly a factor of 2 for methanol and 1½ for ethanol.

This is a very unattractive feature in view of the current need to minimize vehicle size in order to attain the mandated fuel economy standards.

Governmental Incentives

In my judgment, none of the problems outlined above are insurmountable. In general, Ford Motor Co. supports consideration of possible governmental actions which would provide incentive for the development of alternative energy sources. However, we would urge that such proposals provide sufficient flexibility to exploit whichever technology is most effective in meeting the overall energy goals of this country.

For example, we believe that equal consideration should be given to the use of any alcohol, be it methanol, ethanol or other. Which alcohol should be favored is not yet known.

Similarly, if a governmental incentive is found desirable, it should apply equally to the direct use of alcohol and to the use of alcohol after conversion to gasoline. Recent reports indicate that alcohol can be converted to high quality gasoline efficiently and at low cost.

Any proposed incentives should also apply equally to the use of alcohol produced from sources such as grain, sugarcane, and beets as well as to liquid fuels produced from cellulose and other materials present in agricultural and forestry wastes, and from any other currently unexploited source.

Finally, the incentives should apply equally to fuels produced from solar energy, both by the agricultural product route and by other means such as nonbiological photochemical reactions.

This flexibility is essential in view of the unsolved problems and open issues connected with the prospective use of alcohols and alcohol-gasoline blends as automotive fuels including, for instance, potential adverse impact on air quality; potential driveability problems; and potential corrosion and other material compatibility problems.

As I stated in my opening remarks, although probably not insurmountable, these problems are sufficiently serious to demand careful consideration.

In closing, I express my support for the examination of possible means to encourage the production of liquid fuels from sources other than petroleum, but I urge you to broaden any legislative proposals so as to apply not only to gasohol, but also to all other liquid fuels which could be made from abundant domestic resources.

General Motors Co.
Statement Of Joseph M. Colucci

Mr. COLUCCI. Good afternoon, Mr. Chairman. My name is Joseph M. Colucci, and I am head of the fuels and lubricants department of the General Motors Research Laboratories at Warren, Mich. Our department does most of General Motors' research on automotive fuels, including alcohols.

GM's research into alcohol fuels extends back over 40 years. We have experimented in the laboratory with single-cylinder test engines and full-sized production engines and on the road with full-scale vehicles. It is our responsibility to fully understand the operating characteristics of all fuels that are likely to be used in our products. Our primary purpose in appearing here today is to be responsive to your request and share the results of our research work.

At the outset, I should explain that our research effort has attempted to cover use of alcohol fuels, both pure and blended with gasoline. We have found there are advantages and disadvantages to each. Therefore, I shall try to answer your questions first, by using the data and experience from testing 100 percent alcohol fuels as a base of reference, and second, using a 90 percent gasoline to 10 percent alcohol fuel blend.

Extensive Research Into Methanol Use

While we have done extensive research on both methanol and ethanol, most of our experimental work has been with methanol—both as a pure fuel and as a blended fuel. We have done this for two reasons. First, as a simple matter of research efficiency it makes sense to concentrate our effort on one fuel rather than two, when the two fuels have similar operating characteristics as an automotive fuel.

Second, in the long run General Motors believes that methanol is more likely than ethanol to be produced in quantities sufficient to be used as an automotive fuel. I'll say a little more about this later.

In answering the committee's questions, I'll indicate when there is a different answer for ethanol than for methanol. But first let me discuss some of the different properties of these two alcohol fuels.

Differences Between Methanol And Ethanol

Differences between methanol and ethanol: Although ethanol and methanol are distinct chemically, both are clear liquids that look and smell alike. Both also have very different toxic effects. For example, methanol is poisonous and long exposure to its vapor causes loss of vision, while the intake or inhalation of ethanol vapor causes intoxication. Extreme care would have to be taken in handling methanol to avoid inhalation or skin contact. For example, it would be very dangerous to use methanol as a degreasing agent.

Gasoline is now frequently misused as a degreasing agent in home workshops, but it is not as harmful to the skin or to inhale.

Thus, the toxicity of the alcohols must be thoroughly investigated before they gain widespread use as automotive fuels.

Ethanol, the alcohol in beverages, is currently produced by fermentation of grains, while methanol, so-called wood alcohol, is currently produced from natural gas. Both ethanol and methanol can be produced from a wide variety of raw materials. Ethanol is completely soluble in gasoline, while methanol is less soluble.

From here I will highlight portions of the full statement which we have already distributed to you, and we wish to have it included in the official record.

Answers To Questions Directed At GM

With those preliminary observations, I'll now answer the 10 questions directed to General Motors.

Question 1. Are the current automobile components compatible with alcohol fuels, or would some modification(s) be necessary to introduce their use in current automobiles?

Answer. To operate on 100 percent methanol.

It is impossible to operate cars currently on the road with pure methanol because they are neither equipped with the proper carburetors, nor do they have means to vaporize enough fuel to get the car started at temperatures below 45 degrees Fahrenheit. Methanol is more corrosive than gasoline to some current fuel system materials; and specially formulated plastic, rubber, and metal linings would have to be developed for use in the fuel system to resist chemical attack.

For example, the lead alloy coating that lines all metal gasoline tanks can be readily corroded by methanol. Therefore, retrofit of used cars to operate on pure methanol is not practical.

All of these problems, however, appear solvable for cars specifically designed to use methanol. Necessary design changes are shown on the following page.

One of the main problems, cold starting, arises because the heat of vaporization for the alcohols is considerably greater than for gasoline. On an equivalent fuel energy basis, it takes about 8 times more heating of the intake mixture with methanol than with gasoline, and about 4 times more with ethanol. This problem may be overcome by adding volatile materials, such as gasoline, to the pure alcohol to provide the vapors required for starting the engine. Also, intake system design changes could help.

To operate on a blend of 90 percent gasoline, 10 percent methanol:

Some of the same material changes required for use of 100 percent methanol would be desirable for use of the 90 percent-10 percent blend. However, retrofit would not be required absolutely in used cars. A potential, gradual corrosive attack on the fuel system parts would take place, and drivers would experience some degradation in driveability. To retain approximately the same driveability, larger carburetor jets would have to be installed, costing about $35 to $75. The larger jets are required to increase the volumetric fuel flow rate to overcome the leaning effect due to addition of the alcohol to the gasoline. Leaner carburetion generally results in poorer driveability.

How would the above conclusions be different for ethanol? To operate on 100 percent ethanol:

Modification of the fuel system for use of 100 percent ethanol would be approximately the same as those changes illustrated on page four for methanol. In short, cars now on the road cannot operate on 100 percent pure ethanol just as they cannot operate on 100 percent pure methanol.

To operate on a blend of 90 percent gasoline, 10 percent ethanol:

Cars now in use can operate on this blend of ethanol with no adjustment. However, as noted above, a corrosion problem would begin and drivers would experience some loss of driveability. However, the effects with 10 percent ethanol will not be as pronounced as with 10 percent methanol because the leaning effect of the ethanol is less than that for methanol. Problems of vapor lock and the tendency of the fuel to separate will be present with ethanol, but to a lesser extent than with methanol. Generally, ethanol at the same concentrations is easier to accommodate than methanol as a blended fuel.

Performance Problems In Alcohol Fuels

Question 2. Are any performance problems indicated were alcohol fuels to be introduced on a national scale?

Answer. Assuming that this question is directed solely to the driveability aspects of performance, the outline below suggests our experience with methanol, both as a pure fuel and as a blend with gasoline.

The use of ethanol would present generally the same problems, but, again, to a lesser extent than with methanol.

Modifications To Avoid Performance Problems

Question 3. What modifications would be necessary to avoid these problems?

Answer. As indicated in the answer to Question 1, for 100 percent alcohol fuels, new cars would require significant redesign, both for methanol and ethanol use. Cars now on the road cannot use these pure fuels. With redesigned new cars, the problems associated with use of pure alcohol fuels can probably be overcome, especially if the fuel is modified to help solve the cold starting problem.

To avoid driveability problems when using alcohol/gasoline blends in existing cars, two things would have to be done. First, the carburetor has to be modified, fuel orifice size increased, to avoid the problems of surge, hesitation and stumble. Second, to avoid vapor-lock, the fuel would have to be modified to reduce its volatility. This could be done by removing some of the volatile gasoline components. However, this could negate the presumed energy savings obtained by adding alcohol to gasoline.

If new cars were to be designed to operate on alcohol gasoline blends, their carburetor and fuel system materials could be appropriately modified.

Modification Costs

Question 4. How much would you estimate any such modifications would cost?

Answer. GM has not estimated the cost to redesign vehicles to operate on pure alcohol. The need has not arisen for such estimates nor has our development work proceeded far enough to permit us to make them.

For vehicles in use to operate on alcohol gasoline blends, as I mentioned previously, alterations to the carburetor could probably be purchased for about the same cost as a carburetor overhaul—between $35 and $75. If alterations were to be made at the factory, so that new cars would be able to run on a gasoline/alcohol blend, the costs would be negligible.

Required Lead Time To Develop Modifications

Question 5. How much lead time would the industry need to make these modifications?

Answer. There would probably be enough lead time to redesign new vehicles to operate on pure alcohol fuels while the production facilities are being constructed to provide a significant volume of alcohol fuels for national use. The major problem in moving to an entirely new fuel for the national fleet of new cars in any one model year is the availability of a sufficient volume of the new fuel nationwide to implement the changeover—even on a phase-in basis. Such an increase in national production and distribution of alcohol fuels is likely to take several years.

To change the specifications for new car engines so that they will run on either a 10-percent blend of alcohol fuels or straight gasoline, obviously would not take as much time. But, we face problems of another sort before proceeding in that direction. These have to do with certification of new models under the Clean Air Act and compliance with fuel economy standards under the Energy Policy and Conservation Act (EPCA). In our judgment, both the statutes and administrative regulations under these two acts would have to be modified:

One, to meet Clean Air Act requirements we are required to certify our prototype vehicles for sale after complying with procedures which include a 50,000 mile durability test using a special fuel, the content of which is specified by EPA. This certification fuel is supposedly representative of average fuels obtainable in the market place nationwide. Thus, we could not begin to modify engine specifications of our new vehicles to permit use of a blend of alcohol and gasoline until after it was available nationwide at gasoline stations so that EPA would be required to include it as a specification in the certification fuel.

Two, future emission standards are set at such stringent levels that we are having great difficulty complying with the law using a fuel with which we have had over 10 years experience. To require us to use a new fuel would introduce additional uncertainty concerning compliance with future standards.

Three, since pure alcohol fuels deliver considerably less miles per gallon fuel economy—about one-half for methanol, two-thirds for ethanol—and consume significant amounts of energy in processing, serious questions would arise about our capability of complying with the fuel economy standards.

Alcohol Fuels As Petroleum Extenders

Question 6. What are your thoughts on introducing alcohol fuels as petroleum extenders?

Answer. General Motors believes the best use of alcohol fuels, especially methanol, may be with industrial and utility gas turbine engines used to generate electricity. These engines currently use liquid hydrocarbon fuels which can be diverted to transportation fuel use and significantly add to gasoline supplies.

At the same time, industrial and utility use of methanol should not create any major new problems. Instead, it would help these stationary sources of pollution to meet the oxides of nitrogen emission standards, and it would simplify the serious problems associated with distributing alcohol fuels from the refiner to the user—especially if they are to be distributed as blends. It would also obviate the extensive automotive engine and fuel system changes required for operation with pure alcohol fuels—or the engine adjustments necessary for the best accommodation of blended fuels.

Serious distribution and storage problems could be expected for a 90 percent-10 percent blend of gasoline and methanol, because of its tendency, in the presence of even very small amounts of water, to separate into its two components. Separation in the vehicle fuel tank would cause serious problems because the engine would not be able to operate on the pure alcohol which would be at the bottom of the fuel tank.

However, the most substantial problems resulting from use of alcohol as an automotive fuel are those of adequate supply and distribution and the economics of developing this capability. Additional key questions include what feedstocks to use and whether, in the long run, we should convert our nonpetroleum resources to alcohol fuels or to hydrocarbon fuels such as gasoline and diesel fuel.

Fuel Blends Vs. 100 Percent Alcohol Fuel

Question 7. From the automobile industry's viewpoint, would 10 percent blends nationwide be preferable, or would use of 100 percent alcohol fuels in selected vehicle fleets make more sense?

Answer. Our test data and experience leads us to believe that use of 100 percent alcohol fuels is the way to go if alcohols are to be used as automotive fuels.

However, our main concern is over the equally important question of whether our alternate fuel technology should be used to produce gasoline or methanol. Technology is either available, or being developed, to produce gasoline from nonpetroleum resources such as coal and oil shale which the United States has in abundance. If gasoline is produced, it would eliminate two major disadvantages with alcohol fuels:

One, creating a new distribution system capable of overcoming fuel separation problems and the need for additional pumps, tanks, etc.; and

Two, making all the changes to vehicles, either new or used, to accommodate a new fuel with the least degradation of emissions, fuel economy, and driveability.

From a national energy conservation viewpoint, the amount of energy required to produce the automotive fuel becomes important. Coal can be converted to either alcohols or hydrocarbon fuels. The conversion of coal to methanol consumes about 50 percent of the energy in coal, while the conversion of coal to synthetic gasoline consumes less energy—about 35 percent to 45 percent. This may prove to be the most viable and economic route to follow.

Effect Of Alcohol Fuels On Auto Emissions And Fuel Economy

Question 8. How would introduction of alcohol fuels affect auto emissions and fuel economy?

Answer. These results are from tests with present hardware. Data from three-way catalyst hardware does not yet exist. However, it is expected that the results would be about the same.

Costs Of Methanol Ethanol Blends

Question 9. What are the relative costs and benefits of ethanol and methanol blends, or some combination of them?

Answer. The answer to question on costs should most appropriately come from witnesses representing those who produce alcohol fuels. Since General Motors has no expertise in this area, we have no answer for this part of the question.

Information on the various benefits coming from blends of ethanol and methanol, has been supplied in answers to previous questions.

Warranties On Autos Using Alcohol Blends

Question 10. Will current warranties cover automobiles using alcohol blends?

Answer. In general, the coverage of GM's new vehicle warranties would not be voided by the use of an alcohol gasoline fuel blend in a vehicle designed to use unleaded gasoline. However, if the use of an alcohol fuel blend caused a failure for example, in the carburetor because of the corrosive attack of alcohol on metal, it is likely that General Motors would deny that particular claim under the warranty. Other subsequent claims during the warranty period would be considered on their particular merits and may or may not be accepted.

General observations. I would like to conclude by making some general observations about alcohol fuels. My comments here might also serve in part as a summary of our answers to the 10 questions.

Use of alcohol as fuel in automobiles is technically feasible although the result in the case of an alcohol gasoline blend would be some degradation in driveability.

Methanol appears to us to be the most practical alcohol fuel for automotive use in the United States, since it can be readily produced from coal, of which the United States has an abundant supply. Brazil, on the other hand, has operated some of its highway transportation system on a blend of up to 30 percent ethanol in gasoline, depending on its changing market surpluses of the feedstock it uses, sugarcane.

Production capacity for widespread automotive use of either alcohol fuel could not be available for many years. Since the costs of producing these fuels are presently much higher than those of producing gasoline, and gasoline is still available, it is unlikely that the capacity will be developed in response to market demands. However, this situation could change if the national energy situation made use of such fuels either attractive or necessary. If the price of gasoline increased to where alcohol fuels became competitive, the production capacity would likely follow.

Development of a production and distribution system for alcohol fuels would take a large investment and long lead times. During this lead time period any necessary vehicle modifications could be accomplished.

Clearly, alcohols could be used as a strategy to reduce petroleum consumption in the United States. But questions of availability and usage priorities are of paramount importance.

Prepared Statement

This concludes my formal statement. Thank you. I would like to have my prepared statement placed in the record.

[The statement follows:]

Prepared Statement of Joseph M. Colucci
Head of Fuels and Lubricants Department
General Motors Research Laboratories

Good afternoon, Mr. Chairman, my name is Joseph M. Colucci, I am head of the Fuels and Lubricants Department of the General Motors Research Laboratories at Warren, Michigan. Our department does most of General Motors research on automotive fuels, including alcohols.

Introduction

GM's research into alcohol fuels extends back over forty years. We have experimented in the laboratory with single-cylinder test engines and full-sized production engines and on the road with full scale vehicles. It is our responsibility to fully understand the operating characteristics of all fuels that are likely to be used in our products. Our primary purpose in appearing here today is to be responsive to your request and share the results of our research work.

At the outset, I should explain that our research effort has attempted to cover use of alcohol fuels, both pure and blended with gasoline. We have found there are advantages and disadvantages to each. Therefore, I shall try to answer your questions first, by using the data and experience from testing 100% alcohol fuels as a base of reference, and secondly using a 90% gasoline/10% alcohol fuel blend.

While we have done extensive research on both methanol and ethanol, most of our experimental work has been with methanol—both as a pure fuel and as a blended fuel. We have done this for two reasons. First, as a simple matter of research efficiency it makes sense to concentrate our effort on one fuel rather than two, when the two fuels have similar operating characteristics as an automotive fuel. Second, in the long run General Motors believes that methanol is more likely than ethanol to be produced in quantities sufficient to be used as an automotive fuel. I'll say a little more about this later.

In answering the Committee's questions, I'll indicate when there is a different answer for ethanol than for methanol. But first let me discuss some of the different properties of these two alcohol fuels.

Differences Between Methanol and Ethanol

Although ethanol and methanol are distinct chemically, both are clear liquids that look and smell alike. Both also have very different toxic effects. For example, methanol is poisonous and long exposure to its vapor causes loss of vision, while the intake or inhalation of ethanol vapor causes intoxication. Extreme care would have to be taken in handling methanol to avoid inhalation or skin contact. For example, it would be very dangerous to use methanol as a degreasing agent. Gasoline is now frequently misused as a degreasing agent in home workshops, but it is not as harmful to the skin or to inhale. Thus, the toxicity of the alcohols must be thoroughly investigated before they gain widespread use as automotive fuels.

Ethanol, the alcohol in beverages, is currently produced by fermentation of grains, while methanol, so-called wood alcohol, is currently produced from natural gas. Both ethanol and methanol can be produced from a wide variety of raw materials. Ethanol is completely soluble in gasoline, while methanol is less soluble.

With those preliminary observations, I'll now answer the 10 questions directed to General Motors.

Question 1

Are the current automobile components compatible with alcohol fuels, or would some modification(s) be necessary to introduce their use in current automobiles?

Answer: To Operate on 100% Methanol

It is impossible to operate cars currently on the road with pure methanol because they are neither equipped with the proper carburetors, nor do they have means to vaporize enough fuel to get the car started at temperatures below 45 degrees F. Methanol is more corrosive than gasoline to some current fuel system materials; and specially formulated plastic, rubber and metal linings would have to be developed for use in the fuel system to resist chemical attack. For example, the lead alloy coating that lines all metal gasoline tanks can be readily corroded by methanol. Therefore, retrofit of used cars to operate on pure methanol is not practical.

All of these problems, however, appear solvable for cars specifically designed to use methanol. Necessary design changes are shown in Fig. B-17.

One of the main problems, cold starting, arises because the heat of vaporization for the alcohols is considerably greater than for gasoline. On an equivalent fuel energy basis, it takes about eight times more heating of the intake mixture with methanol than with gasoline, and about four times more with ethanol. This problem may be overcome by adding volatile materials, such as gasoline, to the pure alcohol to provide the vapors required for starting the engine. Also, intake system design changes could help.

To Operate On A Blend Of 90% Gasoline, 10% Methanol

Some of the same material changes required for use of 100% methanol would be desirable for use of the 90%/10% blend. However, retrofit would not be required absolutely in used cars. A potential, gradual corrosive attack on the fuel system parts would take place, and drivers would experience some degradation in driveability. To retain approximately the same driveability, larger carburetor jets would have to be installed, costing about $35 to $75. The larger jets are required to increase the volumetric fuel flow rate to overcome the leaning effect due to addition of the alcohol to the gasoline. Leaner carburetion generally results in poorer driveability.

How Would The Above Conclusions Be Different For Ethanol?

To Operate On 100% Ethanol

Modification of the fuel system for use of 100% ethanol would be approximately the same as those changes illustrated on page 4 for methanol. In short, cars now on the road cannot operate on 100% pure ethanol just as they cannot operate on 100% pure methanol.

To Operate On A Blend Of 90% Gasoline, 10% Ethanol

Cars now in use can operate on this blend of ethanol with no adjustment. However, as noted above, a corrosion problem would begin and drivers would experience some loss of driveability. However, the effects with 10% ethanol will not be as pronounced as with 10% methanol because the leaning effect of the ethanol is less than that for methanol. Problems of vapor lock and the tendency of the fuel to separate will be present with ethanol, but to a *lesser* extent than with methanol. Generally, ethanol at the same concentrations is easier to accommodate than methanol as a blended fuel.

Question 2

Are any performance problems indicated were alcohol fuels to be introduced on a national scale?

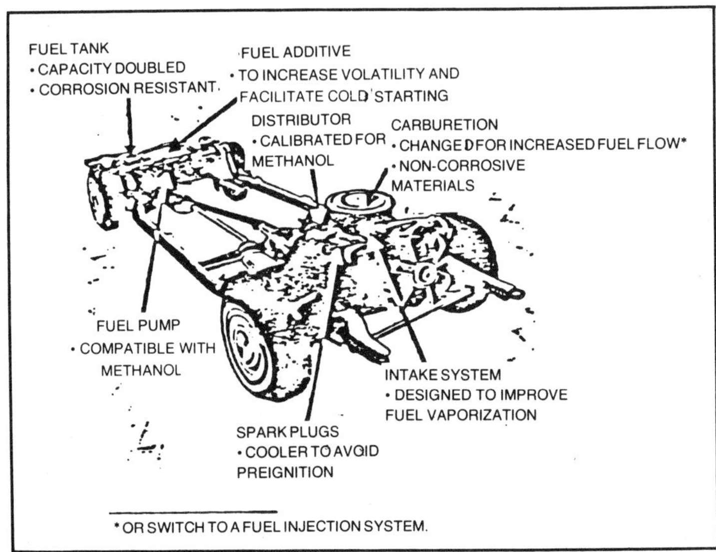

Fig. B-17. Design changes.

Answer:

Assuming that this question is directed solely to the driveability aspects of performance, the outline below suggests our experience with methanol, both as a pure fuel and as a blend with gasoline:

Driveability	90%/10% Blend (No Change to Carburetor)	Pure Methanol (Redesigned Car)
1. Hesitation, stumble, surge	Worse	About the same
2. Vapor lock	Potential problem	No problem
3. Cold Start	No problem	Problem-probably can be eliminated with hardware or fuel modifications

The use of ethanol would present generally the same problems, but, again, to a lesser extent than with methanol.

Question 3

What modifications would be necessary to avoid these problems?

Answer:

As indicated in the answer to Question 1, for 100% alcohol fuels, new cars would require significant redesign, both for methanol and ethanol use. Cars now on the road cannot use these pure fuels. With redesigned new cars, the problems associated with use of pure alcohol fuels can probably be overcome, especially if the fuel is modified to help solve the cold starting problem.

To avoid driveability problems when using alcohol/gasoline blends in existing cars, two things would have to be done. First, the carburetor has to be modified (Fuel orifice size increased) to avoid hesitation and stumble. Second to avoid vapor-lock the fuel would have to be modified to reduce its volatility. This could be done by removing some of the volatile gasoline components. However, this could negate the presumed energy savings obtained by adding alcohol to gasoline.

If new cars were to be designed to operate on alcohol gasoline blends, their carburetor and fuel system materials could be appropriately modified.

Question 4

How much would you estimate any such modifications would cost?

Answer

GM has not estimated the cost to redesign vehicles to operate on pure alcohol. The need has not arisen for such estimates nor has our development work proceeded far enough to permit us to make them.

For vehicles in use to operate on alcohol gasoline blends, as I mentioned previously, alterations to the carburetor could probably be purchased for about the same cost as a carburetor overhaul—between \$35 and \$75. If alterations were to be made at the factory, so that new cars would be able to run on a gasoline/alcohol blend, the costs would be negligible.

Question 5

How much lead time would the industry need to make these modifications?

Answer:

There would probably be enough lead time to redesign *new* vehicles to operate on pure alcohol fuels while the production facilities are being constructed to provide a significant volume of alcohol fuels for national use. The major problem in moving to an entirely new fuel for the national fleet of new cars in any one model year is the availability of a sufficient volume of the new fuel nationwide to implement the change-over—even on a phase-in basis. Such an increase in national production and distribution of alcohol fuels is likely to take several years.

To change the specifications for new car engines so that they will run on either a 10% blend of alcohol fuels or straight gasoline, obviously would not take as much time. But, we face problems of another sort before proceeding in that direction. These have to do with certification of new models under the Clean Air Act and compliance with fuel economy standards under the energy Policy and Conservation Act (EPCA). In our judgment, both the statutes and administrative regulations under these two Acts would have to be modified:

(1) To meet Clean Air Act requirements we are required to certify our prototype vehicles for sale after complying with procedures which include a 50,000 mile durability test using a special fuel, *the content of which is specified by EPA*. This certification fuel is supposedly representative of average fuels obtainable in the marketplace nationwide. Thus, we could not begin to modify engine specifications of our new vehicles to permit use of a blend of alcohol and gasoline until *after* it was available nationwide at gasoline stations so that EPA would be required to include it as a specification in the certification fuel.

(2) Future emission standards are set at such stringent levels that we are having great difficulty complying with the law using a fuel with which we've had over ten years experience. To require us to use a new fuel would introduce additional uncertainty concerning compliance with future standards.

(3) Since pure alcohol fuels deliver considerably less miles per gallon fuel economy (about ½ for methanol, 2/3 for ethanol) and consume significant amounts of energy in processing, serious questions would arise about our capability of complying with the fuel economy standards.

Question 6

What are your thoughts on introducing alcohol fuels as petroleum extenders?

Answer:

General Motors believes the best use of alcohol fuels, especially methanol, may be with industrial and utility gas turbine engines used to generate electricity. These engines currently use liquid hydrocarbon fuels which can be diverted to transportation fuel use and significantly add to gasoline supplies.

At the same time, industrial and utility use of methanol should not create any major new problems. Instead, it would help these stationary sources of pollution to meet the oxides of nitrogen emission standards, and it would simplify the serious problems associated with distributing alcohol fuels from the refiner to the user—especially if they are to be distributed as blends. It would also obviate the extensive automotive engine and fuel system changes required for operation with pure alcohol fuels—or the engine adjustments necessary for the best accommodation of blended fuels.

Serious distribution and storage problems could be expected for a 90%/10% blend of gasoline and methanol, because of its tendency, in the presence of even very small amounts of water, to separate into its two components. Separation in the vehicle fuel tank would cause serious problems because the engine would not be able to operate on the pure alcohol which would be at the bottom of the fuel tank.

However, the most substantial problems resulting from use of alcohol as an automotive fuel are those of adequate supply and distribution and the economics of developing this capability. Additional key questions include what feedstocks to use and whether, in the long run, we should convert our non-petroleum resources to alcohol fuels or to hydrocarbon fuels such as gasoline and diesel fuel.

Question 7

From the automobile industry's viewpoint, would 10% blends nationwide be preferable, or would use of 100% alcohol fuels in selected vehicle fleets make more sense?

Answer:

Our test data experience leads us to believe that use of 100% alcohol fuels is the way to go if alcohols are to be used as automotive fuels.

However, our main concern is over the equally important question of whether our alternate fuel technology should be used to produce gasoline or methanol. Technology is either available, or being developed, to produce gasoline from non-petroleum resources such as coal and oil shale which the U.S. has in abundance. If gasoline is produced, it would eliminate two major disadvantages with alcohol fuels: (1) creating a new distribution system capable of overcoming fuel separation problems and the need for additional pumps, tanks, etc.; and, (2) making all the changes to vehicles, either new or used, to accommodate a new fuel with the least degradation of emissions, fuel economy and driveability.

From a national energy conservation viewpoint, the amount of energy required to produce the automotive fuel becomes important. Coal can be converted to either alcohols or hydrocarbon fuels. The conversion of coal to methanol consumes about 50% of the energy in coal, while the conversion of coal to synthetic gasoline consumes less energy—about 35% to 45%. This may prove to be the most viable and economic route to follow.

Question 8

How would introduction of alcohol fuels affect auto emissions and fuel economy?

Answer:

These results are from tests with present hardware. Data from three-way catalyst hardware does not yet exist. However, it is expected that the results would be about the same.

Emissions	90%/10% Gasoline Methanol (no change to carburetor)	Pure Methanol (redesigned car)
1. HC	about the same	considerably less
2. CO	decreased	about the same
3. NOx	slight decrease	considerably less*
4. Aldehydes**	slight increase	great increase
5. unburned methanol**	increase	great increase
6. evaporative	may increase	may decrease
Fuel Economy		
1. Volume Basis (miles per gallon)	Slight decrease	About 50% less***
2. Energy Basis (miles per Btu)****	About the same	Slight increase

*Could meet .4 g/m NOx standard without three-way converter. Much higher aldehyde and methanol emissions could be controlled through use of oxidation catalyst.

**These emissions are not now regulated.

***One major problem of alcohol fuel use in automobiles is the significant decrease in mile per gallon fuel economy with pure alcohols. To maintain constant range, fuel tank capacity would have to be increased.

****Based on the energy content of the fuel in the tank; does not include energy losses in producing the fuel.

Question 9

What are the relative costs and benefits of ethanol and methanol blends, or some combination of them?

Answer:

The answer to question on costs should most appropriately come from witnesses representing those who produce alcohol fuels. Since General Motors has no expertise in this area, we have no answer for this part of the question.

Information on the various benefits coming from blends of ethanol and methanol, has been supplied in answer to previous questions.

Question 10

Will current warranties cover automobiles using alcohol blends?

Answer:

In general, the coverage of GM's new vehicle warranties would not be voided by the use of an alcohol gasoline fuel blend in a vehicle designed to use unleaded gasoline. However, if the use of an alcohol fuel blend caused a failure, for example, in the carburetor because of the corrosive attack of alcohol on metal, it is likely that General Motors would deny *that particular claim* under the warranty. Other subsequent claims during the warranty period would be considered on their particular merits and may or may not be accepted.

General Observations

I would like to conclude by making some general observations about alcohol fuels. My comments here might also serve in part as a summary of our answers to the 10 questions.

Use of alcohol as fuel in automobiles is technically feasible although the result in the case of an alcohol gasoline blend would be some degradation in driveability.

Methanol appears to us to be the most practical alcohol fuel for automotive use in the United States, since it can be readily produced from coal, of which the U.S. has an abundant supply. Brazil, on the other hand, has operated some of its highway transportation system on a blend of up to 30% ethanol in gasoline, depending on its changing market surpluses of the feedstock it uses, sugar cane.

Production capacity for widespread automotive use of either alcohol fuel could not be available for many years. Since the costs of producing these fuels are presently much higher than those of producing gasoline, and gasoline is still available, it is unlikely that the capacity will be developed in response to market demands. However, this situation could change if the national energy situation made use of such fuels either attractive or necessary. If the price of gasoline increased to where alcohol fuels became competitive, the production capacity would likely follow.

Development of a production and distribution system for alcohol fuels would take a large investment and long lead times.

During this lead time period any necessary vehicle modifications could be accomplished.

Clearly, alcohols *could* be used as a strategy to reduce petroleum consumption in the U.S. But questions of *availability* and usage priorities are of paramount importance.

This concludes my formal statement.

Thank you.

Ms. LUBALIN. Thank you. Both of you have been very responsive to our questions. I think there are one or two things the Senator would like to ask. I will leave it to him when he gets back. There is a vote on now. So, Mr. Wise, if you will proceed now. Thank you.

Mobil Oil Co.
Statement Of John J. Wise
Methanol From Synthesis Gas

Mr. WISE. My name is John J. Wise, and I am vice president for planning of Mobil Research and Development Corp.

Today, I would like to tell you about a new process invented by Mobil scientists to convert methanol into high-octane gasoline. Methanol—also known as methyl alcohol or wood alcohol—can be made from synthesis gas, which can be made from coal by commercially proven technology. The Mobil process provides the final link in a new route to high octane gasoline from coal.

The gasoline is 96 octane, unleaded, and chemically similar to gasoline made from crude oil. It contains no sulfur, nitrogen, or other impurities. With this new process, the Nation can draw upon its abundant reserves of coal to provide gasoline, long after crude oil becomes too scarce or too expensive.

We are now furthering the development of this process under a cost-sharing contract with the Department of Energy. It is the only process that converts alcohol into gasoline. We believe it will be cheaper than the only coal liquefaction process in commercial use today, and competitive with those now under development.

Methanol is in effect about half water and half useful energy. What the Mobil process does essentially is to wring out the water and rearrange the remaining hydrogen and carbon atoms into the concentrated, high energy fuel we know as gasoline. Each gallon of gasoline has 95 percent of the energy contained in the 2 gallons of methanol it was made from.

Catalyst Process

Our process is relatively simple, and uses equipment of the general type used routinely in the oil industry. The heart of the process is a novel catalyst which was discovered by our scientists in the course of developing new methods for making petroleum and chemical products from crude oil.

This catalyst has a unique structure that directs the conversion of alcohol almost entirely into high octane gasoline. We estimate that 2 barrels—84 gallons—of gasoline can be made from 1 ton of raw Wyoming subbituminous coal.

We have just completed a successful 3-month run with a test unit at our laboratory in Paulsboro, N.J. This is a 4-barrel-a-day unit. We are now formulating plans to design, build and operate a 100 barrel-a-day pilot plant. This program will require about 4 years.

The final scaleup could be to a commercial-size facility, which could require another 6 years to get on stream. The technology is so well advanced, however, that the lead time could be considerably shortened if necessary.

High Cost Of Coal Gasification

The primary barrier to commercialization is the high cost of gasifying the coal and converting the synthesis gas to methanol. The Mobil process would represent less than 10 percent of the total investment, and the cost of converting methanol to gasoline would be only 5 to 10 cents per gallon of gasoline produced.

But the total cost of the gasoline would be 40 cents to 50 cents per gallon higher than the cost of gasoline made from crude oil at today's prices.

This is a long-range problem, however. We expect that when the technology is needed, which might be 10 to 20 years from now, the economics will be a lot more favorable.

We are often asked why Mobil is interested in converting methanol into gasoline when methanol can be blended with gasoline. We have investigated this question carefully, since we have no desire to develop a technology that will not be needed. Our studies show that using methanol as an automotive fuel presents a number of technical and economic problems that could best be solved by conversion into gasoline.

The distribution and blending costs alone would be higher than the cost of conversion. One reason is that methanol has only half the energy content of gasoline, so twice the amount would have to be stored and transported. Also, alcohols are highly soluble in water, and water is normally present in gasoline pipelines and tanks. So the methanol would have to be delivered separately to the service station in a water-free system, and blended with gasoline at the pump. All these costs would be in addition to the costs of adapting automobiles to use the blend.

The Mobil process could use other forms of alcohol, including ethanol—also known as ethyl alcohol or grain alcohol. Ethanol can be made from corn, wheat, and other carbonaceous materials by fermentation and distillation, but the costs would be much higher than that of methanol and the potential volume if small.

Estimates we have seen indicate that ethanol would cost at least twice as much as methanol on an energy equivalent basis. We estimate that all of the accumulated U.S. grain surplus—2 billion bushels last year—wouldn't make enough net ethanol to supply 1 percent of our gasoline needs.

Advantages Of Coal Use In Conversion To Alcohol Fuels

To sum up, then, I would like to stress these points: One, Mobil is developing a practical process for converting alcohols into high octane gasoline.

Two, the type of alcohol that could be made in greatest volume at lowest cost is methanol from coal.

Three, converting methanol into gasoline would be cheaper to the consumer than using methanol blends.

Four, while gasoline made from coal cannot compete today with gasoline made from crude oil, the economics could be far more favorable in the years to come.

Five, most important, the Nation's abundant supplies of coal can provide gasoline for many years after crude oil resources are depleted.

Prepared Statement

Thank you for your attention. I ask that my complete statement be placed in the record.

(The statement follows:)

Prepared Statement By John J. Wise

Mobil Research On Alcohols And Conversion Of Coal To Gasoline

Mobil Research and Development Corporation is developing a process for converting methanol into high octane gasoline. The process uses a new type catalyst of the ZSM-5 Class that essentially dehydrates methanol, giving gasoline and water as the primary products. Most importantly, this process can be coupled with commercially established technology for converting coal to methanol (Fig. B-18), providing the first new route for making gasoline from coal in over 30 years. The Department of Energy is extending Mobil's work through a series of cost-sharing contracts.

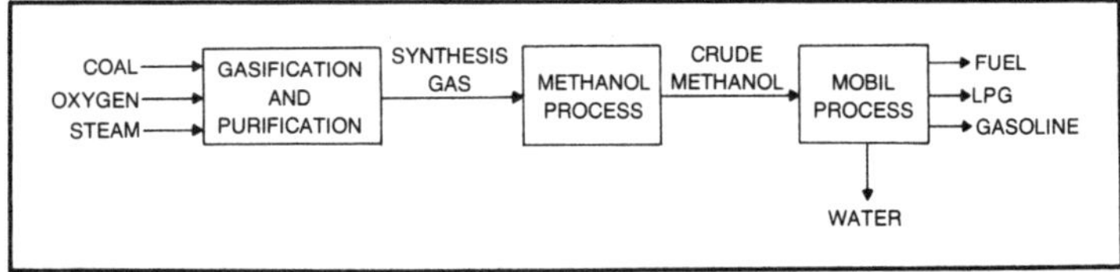

Fig. B-18. Gasoline from coal.

This conversion process is a technological breakthrough, and has significant implications for the future of synthetic fuels, because the U.S. has tremendous coal reserves but not enough crude oil. We believe that Mobil's process will give lower-cost gasoline than the only commercial coal-to-gasoline process (Fischer-Tropsch) and that it could be commercialized more quickly than any other coal liquefaction process now being developed.

This process also can help conserve crude, because each ton of raw, sub-bituminous coal can be converted to about 2 barrels of gasoline, which would otherwise require about 2-½ barrels of crude.

The Mobil process can also be adapted to produce gasoline from ethanol (i.e., grain alcohol). However, we believe that the amount of ethanol that could be made available from fermentable biomass will be very limited. Also, ethanol made by fermentation and distillation will cost more than methanol produced from our abundant coal supplies. Thus, Mobil's process is being tailored to methanol.

Basis Of Mobil's Discovery

Mobil has been developing and commercializing zeolite (crystalline aluminosilicate) cracking catalysts since the 1950's. Currently, more than 95% of all United States petroleum cracking installations use these catalysts. Research on zeolite catalysts has been an increasing effort, in keeping with the ever-broadening potential they have shown for improving petroleum processes and products.

Out of this research came a new family of synthetic zeolites that Mobil calls the ZSM-5 Class of Catalysts. The raw zeolites were found to have a unique channel structure, distinctly different from the familiar wide-pore faujasite and the narrow-pore zeolites. They are selectively penetrated by molecules of intermediate size. In the course of exploring the potential of these catalysts, our researchers found that they could convert oxygen compounds, like methanol, into high-octane gasoline. It has made possible the final step in the coal-syngas-methanol-gasoline sequence.

During the past several years, Mobil has spent about $15 million to develop the catalyst and the methanol-to-gasoline process. In addition, DOE has contributed about $15 million over the past several years, through cost-sharing contracts. We also investigated blending the alcohol with gasoline, but we found this route caused numerous problems and the solution would be costly.

Description

In the Mobil process, methanol is converted to high-octane gasoline using a ZSM-5 Class catalyst. In simple terms, the methanol can be considered to consist of a hydrocarbon-like part and a water-like part. In the Mobil process, the methanol is essentially dehydrated, with gasoline and water as the primary products. This process can be shown schematically as follows:

$$\text{Methanol} \left[\begin{matrix}\text{Hydrocarbon} \\ \text{– Like Part}\end{matrix} + \begin{matrix}\text{Water-Like} \\ \text{Part}\end{matrix}\right] \xrightarrow[\text{Process}]{\text{Mobil}} \text{Gasoline} + \text{Water}$$

or, in chemical terms

$$CH_3OH \quad \left[CH_2 \qquad H_2O\right] \xrightarrow[\text{Process}]{\text{Mobil}} \left[CH_2\right]_n + H_2O$$

In essence, the zeolite catalyst wrings out the water from the methanol, and rearranges the hydrocarbon-part (the carbon and hydrogen atoms) into high octane gasoline. Of course, the detailed chemical reactions are more complex than this simplified description.

The overall material and energy balances of the methanol conversion are illustrated below:

	Methanol →	Hydrocarbons	+ Water
Material Balance:	100 tons	44 tons	+ 56 tons
Energy Balance:	100 BTU	95 BTU	+ 0 BTU

For every 100 tons of methanol converted, 44 tons of hydrocarbons and 56 tons of water are formed. The hydrocarbon product consists primarily of high quality gasoline. The energy balance is extremely favorable; 95% of the thermal energy of the methanol feed is preserved in the hydrocarbon product. Water, of course, has no thermal energy. The remaining 5% goes off as heat of reaction. Thus, this novel conversion of methanol to gasoline allows us to reduce the volume of fuel that must be stored or transported by more than a factor of two, with very little loss in energy of the original methanol.

After the discovery of the novel methanol-to-gasoline conversion, considerable exploratory research was conducted to define the critical parameters. The effects of process variables such as temperature, pressure and space velocity on product yields and catalyst performance were determined.

Based on the promising results of the exploratory research, process development studies were initiated in bench-scale units. A portion of this development work was conducted under DOE Contract No. E(49-18)-1773, which was jointly funded by DOE and Mobil ($1,132,000-70% DOE/ 30% Mobil). The work encompassed process studies in fixed and fluid bed units, characterization of product gasoline, vehicle tests, and engineering design studies.

A long-term catalyst aging test of over 200 days on stream was made in a fixed bed reactor, and process conditions optimized. The technical feasibility of the fluid bed process was demonstrated during a 60-day aging test. The results of the bench-scale tests indicated several advantages for the fluid bed over the fixed bed design. We decided that the next logical step for the fluid bed process was a pilot plant of several barrels per day capacity.

Under another DOE Contract, E(49-18)-2490, a 4 barrel a day fluid bed pilot plant was designed, built, and operated. As before, this contract was jointly funded ($898,000-75% DOE/25% Mobil). Flow studies were conducted in a full-scale, glass reactor to provide data for the reactor design and optimize catalyst efficiency. The results of these flow studies contributed significantly to improving the design and ultimate operation of the pilot plant. The pilot plant was put on-stream in September 1977. Both the start-up and subsequent continuous operation were extremely smooth. During the successful pilot plant operation, it was established that complete methanol conversion and good gasoline selectivity were obtained at design conditions. The catalyst regeneration capability was demonstrated, and the effects of process variables on product yields and properties were determined.

The fluid bed process for the conversion of methanol to gasoline can be explained in terms of the 4 barrel per day pilot plant which is shown schematically in Fig. B-19. Crude methanol is vaporized at the bottom of the reactor, and passes through the dense fluidized catalyst bed at 775°F and 25 psig. The methanol is converted to hydrocarbons and water. The catalyst is removed from reaction products in a disengager section at the top of the reactor, the reactor effluent is condensed, and the water and hydrocarbon products are separated. To make additional gasoline,

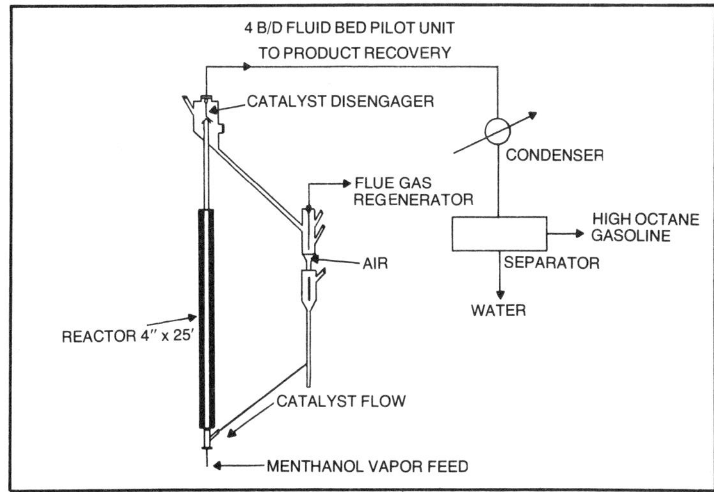

Fig. B-19. Mobil methanol to gasoline process.

propene and butenes can be alkylated with isobutane by conventional petroleum technology.

Portions of the catalyst are periodically removed from the reactor and regenerated with air. The regenerated catalyst is transferred back to the reactor.

Some typical product yields from the fluid bed process are shown in Table B-18. Methanol is converted to hydrocarbons (44%) and water (56%) in expected stoichiometric amounts. Small amounts of carbon monoxide, carbon dioxide, and coke are formed as by-products.

The hydrocarbon product is primarily gasoline. After butenes and propene are alkylated, and gasoline pressured to 9 lb. RVP (Reid Vapor Pressure) with butanes, the finished gasoline has an unleaded Research Octane Number (RON) of 96. The gasoline yield on the basis of total hydrocarbons produced is 88 wt %, and the remaining 12 wt % of light gases are valuable high BTU fuel gas and petrochemicals.

The gasoline produced in the Mobil process is very high quality as shown in Table B-19. The octane values exceed current requirements for either unleaded regular or leaded premium, and the hydrocarbon composition is similar to conventional high quality gasoline. It contains no sulfur nitrogen or other impurities. Its stability (potential gum formation) is acceptable at reasonable additive levels. Durene concentrations are higher than in petroleum-derived fuels; the high melting point of this compound could potentially cause deposits in automotive fuel systems. However, car tests conducted under Contract E(49-18)-1773 with gasolines containing added durene have shown that concentrations up to 4 wt % durene have no significant effects on vehicle performance. It was demonstrated in the 4 B/D pilot plant that durene formation can be controlled below this level by the proper selection of process conditions.

Co-Production With Synthetic Natural Gas

Our description of the Mobil process for manufacture of gasoline from coal-derived methanol has focused on the production of one major product—gasoline. By suitable design of front end units and by judicious use of intermediate products, it is possible to produce comparable quantities of methanol and synthetic natural gas (SNG). Co-production of methanol and SNG from coal is cheaper than the production of methanol alone, and is more thermally efficient because the methane produced in the gasification step is retained rather than converted to additional synthesis gas.

Adding the Mobile process to convert methanol to gasoline results in an efficient process for the production of both SNG and

Table B-18. Yields From Methanol.

Average Bed Temperature, °F	775°F
Pressure, psig	25
Space Velocity (WHSV)	1.0
Yields, wt % of charge	
Methanol + Ether	0.2
Hydrocarbons	43.5
Water	56.0
CO, CO_2	0.1
Coke, Other	0.2
	100.0
Hydrocarbon product, wt %	
Light gas	5.6
Propane	5.9
Propylene	5.0
i-Butane	14.5
n-Butane	1.7
Butenes	7.3
C_5^+ Gasoline	60.0
	100.0
Gasoline (including alkylate) (96 R+0, 9 RVP)	88.0
LP Gas	6.4
Fuel Gas	5.6
	100.0

Table B-19. Typical Properties of Finished Gasoline.

Components, wt %	
Butanes	3.2
Alkylate	28.6
C_5+ synthesized gasoline	68.2
	100.0
Composition, wt %	
Paraffins	56
Olefins	7
Naphthenes	4
Aromatics	33
	100
Research Octane	
Clear	96
Leaded (3 cc TEL/US gal)	102
Reid vapor pressure, psig	9.0
Specific gravity	0.737
Sulfur, wt %	Nil
Nitorgen, wt%	Nil
Durene, wt %	3.8
Corrosion, copper strip	1A
ASTM Distillation, °F	
10%	114
30%	156
504	214
90%	334

gasoline (Fig. B-20). Synthesis gas containing methane would be used as feed to the methanol process. Purge gas from the methanol unit would be methanated to convert unreacted CO and H_2 in the gas to additional methane. The Mobile process would be used to convert crude methanol to gasoline, liquefied petroleum gas and a small amount of additional SNG.

Based on published information, we believe that a co-process based on Lurgi coal gasification, methanol synthesis and the Mobile process could produce two premium products—SNG and high-octane gasoline—with a thermal efficiency of 60-65%.

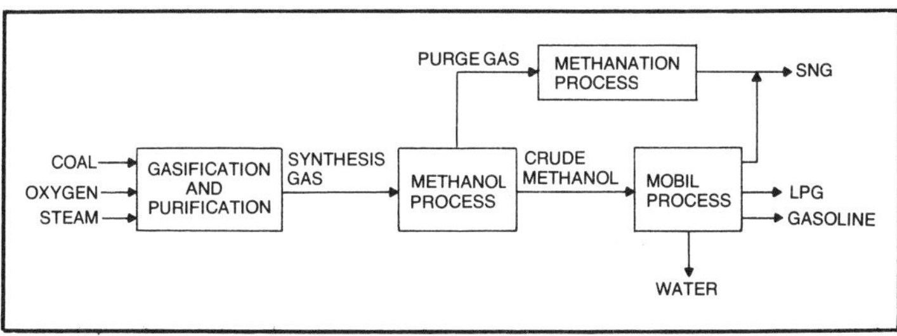

Fig. B-20. Coproduction of gasoline and SNG from coal via methanol.

Economics

The Mobil process would add less than 10% to the investment cost of the coal-syngas-methanol sequence. Most of the cost for the gasoline made from methanol is for making the methanol. We estimate the cost of converting this methanol by the Mobil process at 5 to 10 cents per gallon of gasoline produced. At the current level of methanol-production technology, the cost of gasoline produced by this process would be about 40 cents to 50 cents higher than the cost of gasoline produced from crude oil in the United States today.

Converting coal to gasoline will require very large capital investments. To provide 10% of current domestic gasoline demand from coal would require a capital investment of about $40 billion (1978 dollars).

Alternative Processes

The only coal-to-gasoline technology in commercial operation today is the Fischer-Tropsch process. It is used in South Africa, which has large coal reserves and no domestic oil. But it is a high-cost process that makes low-octane gasoline and lots of by-products.

Compared to Fischer-Tropsch, Mobil's process makes more gasoline of much higher unleaded octane number, less light gas and essentially no fuel oil or oxygenates. Mobil's process also requires only half as many steps to convert synthesis gas into final products. The relative economics of the Fischer-Tropsch and Mobil processes are being assessed by Mobil under a $215,000 DOE contract (ER-77-C-C1-2447), funded by DOE.

We believe Mobil's process will be substantially better than Fischer-Tropsch and at least equivalent to other processes now being developed for making liquid fuels from coal.

Timing

Using the data from the pilot plant, we could design a commercial-size plant. However, we think that good engineering practice dictates that one more scale-up should be carried out. We are now formulating plans to build a 100 barrel-per-day unit.

This pilot plant program, which will take about four years to complete, will provide more accurate data for design of a commercial scale plant. Under normal circumstances, a commercial coal-to-gasoline plant would take about six years to build. The technology is so well advanced, however, that the lead time could be shortened considerably, if necessary.

Alternative Feedstocks

The two common alcohols are methanol and ethanol (grain alcohol). The Mobil process can convert either to the same high octane gasoline. If ethanol could be produced economically, we'd prefer to use it. We chose to concentrate on methanol from coal, however, because we believe it will be cheaper and could be available in much greater quantities. We have seen estimates showing that, on an energy equivalent basis, ethanol would cost twice as much as methanol. Also, the potential ethanol supply is very limited. We estimate that all of the accumulated U.S. grain surplus (2 billion bushels last year) wouldn't make enough *net* ethanol to supply 1% of our gasoline needs.

While ethanol is made only from the sugar and starch components of a biomass, methanol can be made from *any carbon-containing* resource. Methanol could be made from natural gas, shale, tar sand oils, peat, etc. (Fig. B-21). If we wish, methanol can also be made from biomass, with a higher yield per acre than ethanol. But the most economic source, and the only readily available source that could provide very large volumes, is coal. That's why the current Mobil process development, which could be adapted to use any type of alcohol, is presently tailored to coal-derived methanol.

Methanol as Motor Fuel

The use of methanol per se as a blending component of gasoline has been suggested. Methanol is indeed a workable fuel and has advantages in some applications. But, when methanol is blended with gasoline, it poses a number of costly problems. The most troublesome are methanol's low energy density and its high degree of water solubility.

Since a gallon of methanol has only half the energy content of a gallon of gasoline, the volume produced, shipped, and stored would have to be about double that of gasoline.

And since even trace quantities of water lead to phase separation, the distribution system would have to be modified to ensure water exclusion.

Considering its low molecular weight, methanol has a high boiling point. This is due to considerable hydrogen bonding. On the other hand, methanol-gasoline blends have a high vapor pressure. This is because hydrogen bonding between polar methanol molecules is minimized in the non-polar gasoline. As a result, the methanol acquires a vapor-pressure behavior characteristic of its low molecular weight.

To prevent vapor lock problems, butane and possibly pentane would have to be removed from the gasoline. Thus, methanol tends to become a substitute for light hydrocarbon components rather than an extender of gasoline.

Methanol is corrosive to certain metals now used in automobiles. For example, it attacks the lead-tin coating of fuel tanks. The resultant lead hydroxide can plug fuel filters and clog carburetor jets. Methanol also can cause swelling and deterioration of plastics and elastomers. So a number of materials would have to be changed in order to avoid problems for the motorist.

And because methanol is both toxic and miscible with water, any spillage of tank leaks could present unusual toxicological hazards.

It should be emphasized that all these problems can be solved. But just the cost of the special handling procedures to make, distribute and blend methanol would exceed the cost of converting methanol into gasoline. Major costs include:

- ☐ Distributing twice the volume for equal energy content;
- ☐ Modifying the distribution system as needed to avoid water contamination;
- ☐ Isolating methanol from gasoline until it is blended at the pump, and
- ☐ Downgrading butanes displaced because of methanol's high vapor-forming tendency in the blend.

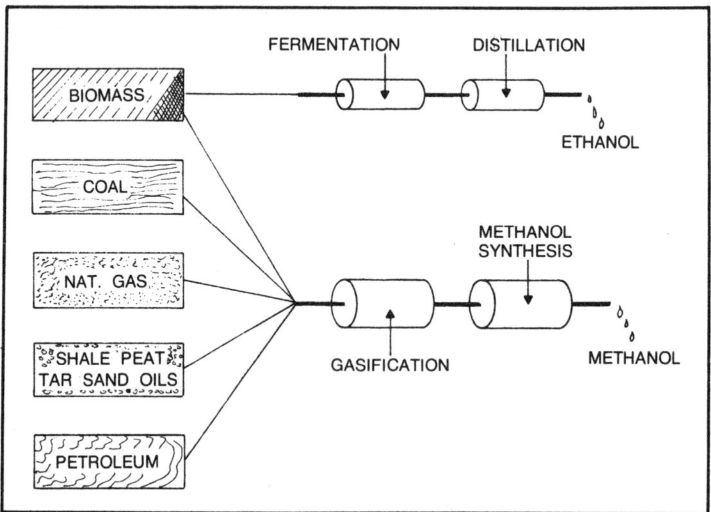

Fig. B-21. Raw materials for ethanol and methanol.

In addition, the consumer would have to bear the cost of replacing methanol-sensitive components in existing cars.

Thus, we believe it would be preferable for many reasons to convert methanol into gasoline rather than blend it with gasoline.

General Questioning

Ms. LUBALIN. Thank you very much, Mr. Wise.

I will substitute for Senator Bayh until he returns. There is a vote going on but he should be back any second now. Mr. Gratch and Mr. Colucci, you were both very responsive to our questions.

Interfacing Of Efforts

One of the things the Senator is interested in how all of these efforts interface or come together. You indicated some modifications might be necessary if blends are to be used in your cars, particularly of materials, and I am interested in getting your thoughts on how these developments interface.

You have indicated that you think once production is sufficient, you will be ready with your cars. I wonder how these efforts mesh. I guess I have heard the figure that, on the average, the American fleet takes 10 years to turn over. If we are interested in providing for the future, we have got to get our cars ready.

I am interested in your thoughts on this question.

Mr. GRATCH. As far as the new cars, there is no question that as soon as gasohol is generally available, we will produce cars that can be used with gasohol. After all, we do produce and sell cars in Brazil that can be used in Brazil with the ethanol gasoline blend supply.

The more difficult problem in the United States is because of the stricter regulations on clean air and the energy policy situation, with the fuel economy. In order for us to be able to do it, either the laws or the regulations would have to be amended to remove some of the obstacles which currently prevent the use of gasohol.

Probably the worse problem is in the cars that were built in the last 6 or 7 years. Prior to 1970, cars had rich carburation and in that case, the addition of alcohol, if anything reduced emissions.

In many of the cars built in the last few years, lean carburation was used and since alcohol could push the carburation to be even leaner, that is what caused some difficulties with the cars in the field. Of course, that is not true in some very specific and relatively rare circumstances such as at high altitude, where the low density of air makes it so that the cars that are designed to be lean at sea level actually are rich. But with that exception, many of the cars built in the last few years could have both higher emissions and driveability problems in operating with gasohol.

Increase In Aldehyde Emissions

Mr. COLUCCI. I would like to add one thing to Dr. Gratch's comments relating to the situation at the high altitude. As Dr. Gratch has mentioned, and I think our testimony also agrees, if you add the alcohol to the gasoline, you increase the aldehyde emissions and these compounds are very photochemically reactive and we should have a very good assessment of their impact on the environment such as Denver's, before we go ahead and try to put gasohol into the use in a plastic Colorado.

The aldehydes could in some way increase the smog instead of decrease it.

State and Federal Elimination of Excise Taxes

Ms. LUBALIN. You apparently have a situation where some States are beginning to repeal or forgive their State excise taxes and there is currently pending in the energy tax bill a forgiveness of the Federal excise tax.

If in fact these efforts come about and people start producing gasohol and selling it in their gas stations, what happens to our fleet of cars?

Mr. GRATCH. As far as new cars, as long as we are allowed by EPA to certify our cars with gasohol, we will be able to design the cars for optimum performance with gasohol.

At present, with the regulations, it forces us to certify the cars with gasoline, a rather well defined set of specifications. Therefore, in order to get our cars certified, we have to design specifically to operate with gasoline and that would make it so that there would be less of an optimum with gasohol.

So there is a regulation problem involved.

Mr. COLUCCI. I agree.

Ms. LAUBLIN. Would that be the particular cue that you would be looking for to start some kind of active plan? Are you advised of the material substitutions and other modifications necessary?

Mr. COLUCCI. I think there would be evidence that there will be in the near future these fuels available on a widespread basis for the public and obviously our efforts then would go on in a parallel fashion to insure our production vehicles are ready to use the gasoline-alcohol blends and they will be out there for the public to buy them.

I don't think it is a major problem coordinating these efforts.

Ms. LUBALIN. I guess I am feeling some concern that there are efforts underway around the country now. They are very, very limited. But if, in fact, something were to happen to change the economics, most likely in the form of a subsidy, or, to look forward, some kind of break through, in 5 or 6 years from now, I am wondering how the industry would respond to those developments to make sure that we have no problems.

Mr. GRATCH. There is no problem, I assure you, the same way that we have had no problem to responding to the changing needs in Brazil. On the basis of earlier testimony, it appears that even most optimistic estimates indicate that from the time one makes a decision, it will take at least 2 years before there is enough gasohol produced to make a difference.

I am quite confident that that will be a sufficient amount of time for us to act, assuming that the regulations are changed so as to allow us to act without violating rules or regulations.

Unleaded Gasolines

Mr. COLUCCI. I think in the past there was a somewhat analogus situation which was the introduction of unleaded gasoline which essentially a new fuel was available and the vehicles were modified to accommodate the new fuel and everything came together at the right time.

So I think a similar kind of situation could occur if it was decided gasoline and alcohol blends or even pure alcohol would be available for public use.

Ms. LUBALIN. My recollection is that unleaded gasoline came on the market in response to air pollution requirements and the kinds of configurations you were designing to meet them. We are talking about the reverse situation now.

Mr. COLUCCI. The point was that the unleaded gasoline had to be widely available in the cars that required it, and the whole thing was coordinated and it worked out fine. I think the same situation would occur here.

Mr. GRATCH. Let me amplify on that. Even though unleaded gasoline was introduced in response to our need to protect catalysts from lead, it did require some vehicle changes. The most important of those was the introduction of converters and we responded to that without any problem.

By the same token, I am quite sure that if gasohol became available and if certification regulations were changed so as to allow us to certify our cars with gasohol, we would respond in kind with no major problems.

Senator BAYH. I apologize. I got just about to this door and then I had to go the other way to vote.

I guess you hit those areas that I am particularly concerned with and I won't ask your indulgence to go over the ground again. I will just read the record.

Warranty Problems

What about the warranty problem? Are current warranties covering cars that use alcohol blends?

Mr. GRATCH. Senator, that is a legal question. As far as I know, none of us is a lawyer, but to the best of the layman's knowledge, while strictly speaking, the use of gasohol would violate the warranty, the practice has been, and I am sure it will continue to be, that unless a failure were identified as being caused by the use of gasohol, these warranties would be honored anyway.

So that the only case of any difficulty would be if there were any corrosion problem, or wear problem that could be attributed to the use of gasohol and then in respect to that, with current cars the warranty would be violated.

Mr. COLUCCI. Senator, the General Motors opinion regarding that question is identical to Mr. Gratch's statement.

Leadtime Question

Senator BAYH. Let me ask Mr. Wise a question or two here. What is your response, sir, to the leadtime question?

Mr. WISE. We are currently at the position now where we completed a 4-barrel-a-day pilot plan, 25 feet high. We think we could even design a commercial size facility to convert methanol into gasoline if we had to.

We think prudent engineering requires one more scale-up step of 100 barrels a day and then we would be ready to design a commercial plant with some reasonable degree of confidence.

A 100-barrel-a-day plant will take us about 4 years to build, operate, get the data we need. Then we would be in the position to build the plant to convert methanol to gasoline. To build the complete facility to go from coal, to methanol to gasoline might under normal circumstances take 6 years to build and put on stream.

Direct Conversion From Methanol To Gasoline

Senator BAYH. You go directly from methanol to gasoline?

Mr. WISE. Yes; essentially, Senator, our argument is that there is no need to put alcohol into gasoline. Our argument is that we have a process that effectively converts alcohols into gasoline with 95 percent thermal efficiency. You reduce the volume by half. You get rid of all the problems you have with distributing alcohols.

We believe the type of alcohol that would be the cheapest is methanol made from coal. I think everyone agrees that methanol is going to be a lot cheaper than ethanol. So we could convert this methanol into gasoline and it would be cheaper to the consumer than the cost of distributing the methanol in the blend.

Ethanol/Methanol Costs

Senator BAYH. May I ask why you think everybody agrees ethanol is going to be more expensive than methanol?

Mr. WISE. I think it is simply due to the fact that coal is a cheaper material than grain or even cellulose. Coal is a relatively cheap raw material and it can be converted into estimates we have seen, show that methanol made from coal may be at least half as expensive as ethanol made from grain, at least half.

I really think the consumer is better off having the lowest cost product. We believe that would be methanol made from coal. We also don't believe there are very—I mean farming is a fairly energy-intensive system, when you look at how much energy it takes to grow the grain, your net production of energy after you subtract what it takes to grow it really isn't all that great.

There is nothing compared to the reserves we have in coal.

Senator BAYH. You know better. You didn't get where you are and you didn't get in this crazy business in the first place by thinking with limited horizons. I will say that as a compliment. Let me finish my question.

Mr. WISE. I am sorry.

Senator BAYH. It might affect the answer. (Laughter.)

Frankly, I don't know what the answer is. We have been told here by our Government officials, that if you look at the difference in the expertise that is available, and the amount of study that has been done on methanol as compared to ethanol, it is like the little league versus the world champs—150 million bucks compared to 6 to 12. Big deal. I wonder, are you prepared to just rest on that now? Don't you think we ought to really move out there? What you said about energy consumption on the farm is true. But if you indeed can come up with a crop or a waste product, reduce the amount of energy used—you have heard our discussion—you could harvest it and be a little more energy efficient—isn't that worth pursuing?

Mr. WISE. Let me say from my point of view, what I would like to get is the cheapest source of alcohol, because in our process, we will convert either alcohol into the same high-octane gasoline.

Senator BAYH. You can convert the same way?

Mr. WISE. The same way. So really as far as I am concerned, it really makes no difference. It is just my opinion from the studies we have done that methanol is going to be cheaper than grain alcohol based on what we have seen. If someone can produce biomass ethanol at a price cheaper than you can make methanol from coal, so much the better. We will convert that gasoline.

Cellulose Process

Senator BAYH. There has been a good deal more work done on methanol than ethanol. We just heard about a process which is relatively new, the cellulose process, which even its fine developer, Dr. Tsao, suggests could be more effective. That is new research and I hope we can move on in that field and explore the horizons in a limitless way.

But let's look at the Mobil process. The current estimates are somewhere between $1.10 and $1.30 a gallon. Do you think with practice and perfection that can be lowered in price?

Mr. WISE. The cost of making gasoline by the Mobil process?

Senator BAYH. Yes.

Mr. WISE. I think that if research is done to cheapen the cost of making methanol or if, indeed, ethanol could be made even cheaper, the Mobil process only represents 10 percent of the total cost of the final product. So it is really the raw material, the alcohol coming to the Mobil process, that represents 90 percent of the cost.

Any research work directed toward cheapening the production of alcohol would benefit this process and the Government has been doing a lot of work on trying to cheapen the gasification step which really is the really big item. It is that coal gasifier, out front, that represents maybe 60 or 70 percent of the total cost. So that I support the work of the Government, the Department of Energy, in trying to cheapen the coal gasification step.

Cost Factors of Petroleum Sources

Senator BAYH. May I just toss out one thought that so far has been borne out, though we don't know what the future is going to hold: That is, one thing we can be certain of is that as the quantity of petroleum gets less, the cost is going to go up. And the price of OPEC oil, even if the quantity doesn't decrease, is going to go up.

There has been a very close relationship, has there not, in the increased cost of coal and the increased cost of petroleum? If you take the escalating cost of petroleum and the escalating cost of the coal and compare that with grain or farm produce, there really hasn't been the same history, has there?

When we talk about 10 years from now, what are the cost factors?

Mr. WISE. I am afraid I am not an expert on what grain prices are and can't comment on that.

Coal Utilization In Process

Senator BAYH. As I said earlier, as enthusiastic as I am about gasohol, or alcohol, or the Mobil process which utilizes coal, I don't want us to develop them at the expense of a foreign commodity price, say, $1.50 for a bushel of corn, over which people go broke. But if you put corn at a reasonable level, it just sort of stays there. It has not gone up significantly. That might be something that we all ought to consider.

I don't know the answer to that. The price of grain might go up. I am not too sure that the price of corn stalks or cobs would go up the same way. All of these have to be cranked in for consideration.

Could all types of coal be utilized in your process?

Mr. WISE. Yes. Any coal you can gasify. We obviously would use the cheapest coal because it turns out the cheapest coal is also the easiest one to gasify.

Elimination of Sulfur and Byproducts

Senator BAYH. Do you get rid of the sulfur problem and byproducts?

Mr. WISE. Yes; sulfur and all the impurities are taken away, scrubbed out. The gasoline we make is pure; no sulfur.

Senator BAYH. Is it cleaner burning gasoline than we have now or would we have the same exhaust problem?

Mr. WISE. It is actually cleaner burning but you would still need the catalytic converter to clean it up.

Investment Costs

Senator BAYH. What does it cost for an economically sized unit to go through that whole process? What does that cost as far as investment is concerned?

Mr. WISE. You are talking very large investments when you go from, say, coal and go all the way down to gasoline. I don't recall the exact numbers, but something on the order of, I guess, 20,000 or 30,000 barrels a day. You are talking over $1 billion.

Comparison of Investment Capital

Senator BAYH. Could I ask you to give a little—this is sort of a quick question and a quick response. I think this is a critical question to consider. Industry and government has only so much capital and this is one of the things you just have to consider, the amount of capital involved. I think it would be one thing that we ought to consider. I don't know what the answer is. You might think about it.

Mr. WISE. We have done some estimating. We estimated that something on the order of replacing 10 percent of the gasoline supply in the United States would be on the order of $40 billion. It is an expensive proposition, compared to refining crude oil today.

Government Assistance In Promoting Fuels Program

Senator BAYH. Let me ask all three of you what I think is a tough question. But I think if we can find the answer, we make it a lot easier to find the answer to these other more technical questions.

What can we in Government do to help create the environment in which we can get those of you in industry who have the technology to join our efforts. The private sector is basically where the necessary combination of technology and experience is. We have some in our institutions of higher learning, in some of the research centers. These places have a great deal of brainpower. But many have not applied what they know to practicalities. What can we do, looking back at our past experience, with gasoline, air emissions, exhaust emissions, the water control experience we have had, air quality in cities—what can we do to create the environment in which we can get business and industry sitting down with those in Government so that the question is not so much what you are going to do but how is the best way of getting there?

We would have been so much better off if we could have found some way where everybody said at the beginning, all right, we are going to have χ particulate matter and NO_x and sulfur dioxide, and the question is not what we are going to do but how we are going to get there. Everybody would have taken the time necessary. We wouldn't have had to worry about the 1970 deadline, or 1971, 1974, or 1975. But we would have determined we were going to take a reasonable amount of time necessary to accomplish the goal. But we were going to reach the goal?

How can we do that? What can I do to be helpful? Does it take mandatory requirements to get people's attention? That is what some of us felt when we looked at emission controls, but we have learned a lot. I think everybody has since then.

Where are we? What is the answer to the question?

Mr. GRATCH. Let me try that. Of course, in some cases the mandatory requirements are the only way. When the customer cannot receive the benefit of any action and when antitrust regulations do not allow us to work jointly, the only way that we can economically put certain things into effect is by mandatory requirements. For instance, when Ford initially introduced seatbelts, we couldn't get people to buy them until the fitting of cars with seatbelts was made mandatory. So certain things have to be mandatory.

Regulation Requirements

Senator BAYH. Just exactly what do you feel? I know that particular situation from a personal standpoint, because back then we had a General Motors product which was involved in a head-on collision, caused by another darned fool. So we replaced our car shortly thereafter by a Ford, because at the time it was the only vehicle that had a padded dash and seatbelts.

But I think for a lot of people who haven't been involved in a crash, it took maybe 5 to 10 years to get them into that position.

Mr. GRATCH. In some cases there is no question that the regulations are required. However, more care should be exerted to avoid regulations that are counterproductive. Let me give you a simple example of what a regulation is that may well prove to be counterproductive.

That is the way in which fuel economy averages have been required in the Energy Policy and Conservation Act. I personally support the concept of the act, but there is a deficiency in some of the details. In order to meet those requirements it may be necessary for the automotive industry to go to vehicles that have a much higher octane requirement than is optimum for the utilization of petroleum.

It so happens that high octane gasoline allows you to do things to the car which give you better miles per gallon, but it takes more petroleum to produce the high octane gasoline. If the fuel economy standards are just based on miles per gallon without any statement as to the other requirements of the gasoline, the net result may actually be a worsening.

Another place that the regulations may be counterproductive is in the case of a person such as a person who needs a 12-passenger car; if the fuel economy regulations make it impractical to build new 12-passenger vans, I will have the choice of either going home to use an old van which doesn't have all the improvements and therefore will not take advantage of the fuel economy improvements or when I take all the kids out to dinner, I take them out in three separate Pintos and the fuel consumption of three Pintos is more than a single wagon. There are a lot of regulations nowadays that are really in my judgment counterproductive.

Requirements Established By Marketplace

Senator BAYH. That is the kind of thing that I think we need to address. In my judgment, you can't say to a fellow who has 10 kids you can't buy a car. But what ought to be required is that however big that van, it ought to be the most efficient vehicle you can produce; not necessarily one that has 400 horses under the hood.

The octane business, maybe so. That is something we are going to have to address. If indeed that is what we have required, it is our error. But I tell you, you must look at the other side of this. Just 2 or 3 years ago, when we tried to get requirements, and I myself tried to get certain increases, the industry said there was no way to meet that goal. Yet we have found consistently that you guys can do darned near anything if required to do so. You have fantastic competence.

The question is how we get you to do it in a way that isn't counterproductive? I am going to have to go vote, which may be counterproductive. (Laughter.)

If you would, for the record, the rest of you, give us your thoughts as to what we need as far as the environment here, does it take longer, does it take mandatory standards; by χ year all cars shall be able to use χ percentage of gasohol? I don't know. I don't like the mandatory business. I would rather do it voluntarily.

We have to reach certain goals. Excuse me.

Mr. COLUCCI. I think in addressing the situation, I think General Motors firmly believes in the free enterprise system and I think we would prefer that the marketplace define the future needs.

The other thing, too, I think we want to be sure if we do make future decisions regarding the new fuel, that it be based on the best possible information which can be used. In this regard, I would encourage anything that the Congress can do to provide an interchange between the people of all the organizations that might be concerned in something like this, to provide the information and help to design the right kind of studies that will provide the information needed to make these eventual decisions.

We are not talking about small decisions. These are major decisions for our Nation and I think we need the best possible information available.

One of the things you are going to have to do with respect to switching to new fuel, at least as far as it involves the use in automobiles, is to run sufficient tests to demonstrate that there are no problems or, if there are problems which are covered, that we can then develop solutions. I think this requires controlled testing, not brand-new haphazard testing.

I think we can do that proper kind of testing by working together, by getting all of the necessary parties involved.

Mr. WISE. We also agree with General Motors that we would like to see the marketplace set the requirements whenever possible. I also agree with Dr. Gratch, that in cases where the Government does feel an obligation to set an overall standard, we would like to make sure that hearings such as these are held so that all of the available information can be brought to bear.

The ideal situation, from our point of view, is to set goals rather than to set specifics on how to do it. I will give you a little example.

If one were to mandate the use of methanol at a 10-percent blend in gasoline today, that would be counterproductive from our point of view because what we would do now is back out butanes which are now used as fuels and have almost the same energy content as the methanol we are putting in and then burn the butane in the boiler somewhere, so that the net effect would be no additional gasoline supply. It would just mean we are burning more butanes. In fact, we could have burned the methanol in the boiler in the first place.

So that there are these kind of subtle counterproductive things that one has to be careful of and we would like to see incentives, goals set broadly.

Tax Reductions For Gasohol

Dr. Gratch mentioned earlier in his testimony that rather than legislate incentives for gasohol by saying there is a 5-cent reduction in the tax for alcohol containing blends, we would like to see that more broader, maybe a 5-percent reduction in the tax for gasolines containing synthetic gasoline, because under the current legislation we would mandate the tax rebate only for gasoline derived containing alcohol to eliminate any consideration from using the Mobil process to convert those alcohols into gasolines which we think would be cheaper to the consumer and eliminate all of these kinds of problems that one has for using alcohol.

Ms. LUBALIN. I know Senator Bayh will want to keep in touch with you. I am not sure he is going to get back. I think the kind of concern the Senator was stating before is that we get into kind of a catch 22 situation where we don't want to impose deadlines which may result in unfortunate consequences, but often if we don't set deadlines we don't get to where we need to go on time.

We will be in close touch with the Department of Energy, particularly on this kind of question. I think there is some feeling on Senator Bayh's part that there has been a tendency to stay in the research and development stage too long, with insufficient attention to separating out technologies that could be commercially viable, or are very close to being so, from those still in the R. & D. stage. There is a tendency to keep them in the laboratory too long.

We certainly appreciate your being here today and will be in touch. Thank you very much.

Additional Material and Prepared Statements

Senator Bayh has asked me to convey to all witnesses that the record will remain open to accommodate additional material.

Conclusion of Hearing

(Whereupon, at 4:04 p.m. Tuesday, January 31, the hearing was concluded, and the subcommittee was recessed to reconvene at the call of the Chair.)

Index

E

F

G

H

I

J

K

L

M